Manufacture of Quality Synthetic Detergents

This book provides a thorough exploration of the step-by-step manufacturing processes utilized in the global production of synthetic detergents. Drawing from real-world manufacturing experience at various detergent plants both domestically and internationally, the text presents valuable insights into the industry.

Manufacture of Quality Synthetic Detergents: Process Chemistry and Unit Operations discusses the essential aspects of detergent processing techniques, quality by design, risk assessment and management, and operational requirements. The book also covers the regulations and guidelines set forth by reputable organizations such as the ACS, AIChE, ICE, and AOCS. In addition, it provides a detailed explanation of quality methodologies for granular detergent size and agglomeration, as well as offers techniques for systematic process optimization and good manufacturing practice. The importance of paying attention to detail in process development and validation is emphasized, with real-world examples showcasing various facets of synthetic detergent manufacturing.

This comprehensive guide serves as a valuable reference for chemists, chemical engineers, manufacturing professionals, and engineers working in granular and liquids sciences, and manufacturing. It is an indispensable resource for those seeking to enhance their knowledge and skills in the field.

Manufacture of Quality Synthetic Detergents

Process Chemistry and Unit Operations

Sam A. Hout

CRC Press
Taylor & Francis Group
Boca Raton London New York

CRC Press is an imprint of the
Taylor & Francis Group, an **informa** business

Designed cover image: Shutterstock

First edition published 2025
by CRC Press
2385 NW Executive Center Drive, Suite 320, Boca Raton FL 33431

and by CRC Press
4 Park Square, Milton Park, Abingdon, Oxon, OX14 4RN

CRC Press is an imprint of Taylor & Francis Group, LLC

ISBN: 978-1-032-97851-2 (hbk)
ISBN: 978-1-032-97938-0 (pbk)
ISBN: 978-1-003-59621-9 (ebk)

DOI: 10.1201/9781003596219

Typeset in Times
by Newgen Publishing UK

Contents

Preface vii
Acknowledgments ix
About the Author x

Introduction 1

Chapter 1 Detergent Raw Materials 13

Chapter 2 Sulfonation and Sulfation 34

Chapter 3 Acid Mix Concentration 51

Chapter 4 Neutralization 56

Chapter 5 Paste and Crutcher Mix Phase Systems 64

Chapter 6 Crutching 70

Chapter 7 Spray Nozzles 85

Chapter 8 Spray Drying 90

Chapter 9 Product Handling 106

Chapter 10 Special Processing Steps 119

Chapter 11 Packing 134

Chapter 12 Finished Product Quality 139

Chapter 13 Behavior of Product in the Market 147

Chapter 14 Material Balances 156

Appendices 161
Index 196

Preface

Synthetic detergent manufacturing engineering is a highly categorized discipline. It is my intent for this book to minimize the detailed chemical mechanisms of heat and mass transfer, mathematical modeling, and several other analytical and numerical methods and focus on fundamentals. Professional chemical engineers, and other professionals who want to understand the natural sciences behind synthetic detergents chemical engineering principles and practice for applied implementations can benefit from this book comprehensive coverage that includes case studies for overall design, installations, and construction of manufacturing unit operations and process chemistry including standard operation procedures.

The more complex the chemical industry becomes, the more vital the role of comprehensive understanding of processing. The transport of fluids in detergent manufacturing and processing is complex through pipes and vessels that require modeling for design conditions and conversion utilizing temperatures, pressures, and fluid flow rates calculations. Extensive mathematical calculations and materials balance equations based on theory and empirical knowledge allow us to build and construct unit operations for processing, separation, and agglomeration of various compounds into granular products. The developments include biological and powder systems for specialized cleaning performance of fabrics. This includes a spectrum of different materials that extend from pure simple solutions to complex emulsions. Cleaning sanitizing agents, as well as synthetic detergents all played a major role in the development of chemical engineering in this sector.

I conceptualized writing this book on manufacturing design and operation applications in the synthetic detergents industry after spending over 30 years in manufacturing engineering. Quality by Design (QbD) dictates process preparations and compounding of formulations are key steps in transferring bulk product batches from beginning to end. These processes all require diligent care and technical know-how in transferring across process steps while maintaining clean and safe operations. Personnel flow and materials flow separations and consideration for movement have rigorous requirements to ensure against cross contamination and making a safe product. During my career, I also worked on projects and consulted in various chemical engineering affiliated industries. I wanted to put it all together in one accessible simplified experience-based reference to reinforce knowledge of design, processes, compliance regulations, and provide an easy guide to follow. This book is important in providing a step-by-step understanding of what is required to engineer and manufacture synthetic detergent products.

Process flow, product risk-based quality approach, and the significance of process validation is emphasized. Critical step-by-step technologies and systems are described as part of the integrated facilities design. Design calculations based on fundamental principles of chemical engineering are categorized to include the use of partial differential equations and computerized methods. Reliability in batch-to-batch consistency and repeatable production in a compliant manner are discussed.

The systematic management of risk to ensure efficacious and safe products and manufacturing systems are explained in terms as it relates to process validation.

The purpose of this book is to help my colleagues in this sector of synthetic detergents manufacturing have a simple reference book on important issues that we deal with daily and are always looking for ways to troubleshoot and solve problems that face us in manufacturing products.

To develop this book, I had to learn more about the subject matter that I am describing and covering. I had to investigate insights and real examples about situations where I encountered difficult and complex formulations that required specific innovative ways to process while ensuring compliance. I believe that through learning, training, and asking questions about alternatives, I was able to grow in how to convey this message in a more concise and focused fashion.

The feedback from the reviewers opened my thinking to how others are perceiving the materials and required me to add value in terms of case studies and real examples from experience. Besides writing from my personal experience in industry, I had to research and follow guidelines among other references that have been reviewed and cited as part of this publication.

I had many communications with colleagues to see how ideas can be explained and described in the desired methods. I have been thinking about how to aid on technical subjects that are pertinent to this field for over 30 years, and I started putting together notes that would help my colleagues in the industry for over 20 years.

The table of contents in this book flows as intended for the reader to explore subject matter in a continuous fashion or refer to a specific section as needed. Engineering principles, procedures, and calculations are the foundation of this book to cover overall facilities adherence to SOP and organizational structure, including appropriate protocols and reports covering all aspect of a product submission and process design for pre-approval, namely, exhibit data, and process step-by-step calculations—all presented in an orderly manner with full documentation of real examples from industrial cases.

Acknowledgments

This book is written with support of my family, my caring loving wife Mona, my empathetic daughter Samantha, and my wise son Owen. In addition, to the memory of my beloved brother Dr. Salim Hout, M.D., who helped form my rational approach to thinking and motivated me with unconfinable support, and gratitude to my dear cousin Dr. Nael Hout, PhD, for knowledge management, who helped explaining many complex solutions to problems during our upbringing and pursuit of knowledge. I am indebted to my brotherly friend Sarkis Khawaja for his everlasting support, generosity, and wise guidance. Modestly, many colleagues and friends across the globe have played a transformational role in adding to my knowledge and experience to add value in what I learned and how to share in the service of others. My gratitude for colleagues and friends who contributed to my project management, engineering and managerial execution capabilities, manufacturing science, and technology through our many communications over three decades. Special dedication is hereby expressed to my colleagues at the University of Bath, UK, who advanced my professional knowledge in advanced chemical engineering. In addition, my gratitude goes to my extended family for their respect, moral support, and dedication to our family traditions that emphasize care and compassion to others.

About the Author

Sam A. Hout, Ph.D., MBA, is a Chartered Chemical Engineer and an expert in process engineering. His experience spans over three decades in design, construction, and validation of manufacturing and delivery systems in commercial applications. As a senior consultant, he provided design, installation, construction, and inspection services both in the US domestic and international markets. He was a senior director of engineering, project management, and process technology transfer at Siegfried, Irvine, CA. He held engineering and operations management positions at Teva pharmaceuticals, Johnson & Johnson, P&G, and US government process chemistry research with emphasis on data integrity. For the past 30 years, his focus has been to design, validate, and operate manufacturing facilities with automated capabilities.

Introduction

Formulating detergents to cope with lowered washing temperatures and the increasing share of the wash load held by synthetic-fiber textiles while handling the heavier soil loads is a common approach with European markets. More widespread has been the need to reformulate, or at least re-evaluate, builder systems, as actions against phosphates continue in Europe and are rekindling in the U.S. after nearly decades of calm. With the reformulations, the soap and detergent industry offers yet another example that in new directions, it is the smaller company that takes the chances and makes the new-product introductions. Given an industry as vast as soap and detergents, it probably is not surprising that it tends to be dominated on a world scale by a handful of companies. Looking strictly at soap and detergent sales, at the top of the pile is Procter & Gamble of Cincinnati. London-based Unilever is second. New York City-based Colgate Palmolive is third, and Henkel of Dusseldorf, Germany, is fourth. Unlike the U.S. market, however, where P&G, Colgate, Lever, and the U.S. fourth, Purex Industries totally dominate the business, on a world-wide basis, the international Big Four have less than half the market. The remaining share, about 53 to 55%, is split among innumerable small, local, or regional producers. In Europe, for example, the Association Internationale de la Savonnerie et de la Detergence, an association of European national associations in the industry, estimates that its member groups represent more than 700 detergent and soap makers. Moreover, on a worldwide basis, the Big Four generally tend to do most of their business close to home. P&G's total sales, for example, are about 70% in the U.S. Unilever's are about 65% in Europe, and Henkel does about 80% of its business in Europe. Only Colgate has less than half its sales at home: about 40% in the U.S., and the rest split almost equally between Europe and the rest of the world. The situation is somewhat different with raw material suppliers. There are a few small local or regional suppliers, but basically major national or multinational companies supply most detergent active ingredients, or formulating products such as builders, bleaches, and fillers. The major detergent producers provide such a large demand that suppliers have to be large to produce enough material to supply their big customers. Some companies have divisions to concentrate strictly on this market. BASF, for example, has just restructured some of its operations into specialty chemicals, for example, detergents division. Because most of the active components in modem detergents

DOI: 10.1201/9781003596219-1

are petrochemical-based, suppliers have brought along the economies of scale that always have accompanied basic petrochemical production. Perhaps the trend having the most profound impact now on the soap and detergent industry worldwide is a move toward lower washing temperatures. This trend has been taking hold for a number of years in the U.S., of course, particularly as energy costs escalate. But in the U.S., "hot" has always been a relative term: probably about 50 °C to 60 °C. In Europe and many other areas of the world, "hot" is 90 °C to 95 °C, often referred to as "at the boil". Bleaching systems were formulated originally for optimum performance at these temperatures, and the white cottons and linens used then came out sparkling clean. Energy-consciousness has hit the European consumer hard, however, and both detergent makers and washing-machine producers have been working to design systems that will do the bulk of a family's wash load at 60 °C, with an increasing option of doing some of the wash at an even lower temperature, usually about 40 °C. In that, the industry is approaching the level of that in Japan, where the usual laundry temperature is 30 °C or below. In Mexico, an even more extreme example, historically laundry has been done in cold water.

Temperatures as low as that pose challenges for detergent formulators whose palette up to now has been filled with materials that perform best in warmer water. Formulators have to make sure the active will dissolve readily in cold water. One commonly used surfactant in the U.S., for example, tallow alcohol in bleach system, will perform efficiently. However, perborate bleaches must have boosters or activators when put in products destined for cool- to warm-water washing, or else lose much of their efficiency. They have to make sure the foam generated will be sufficient to assure the consumer that the detergent is working hard, but not so much that it overflows or clogs up a system.

Most detergents on the market are still aimed at the warm wash, although many packages, especially in the U.S., do note that the products are suited for cold-water washing. Lever, of course, sells cold-water all in the U.S.

Heavy-duty liquids formulation has been quite popular in the U.S. detergent market, capturing just under 40%. European researchers contend that the liquids simply are not so effective as heavy-duty powders, and that whereas the difference in performance might not be so noticeable on the lighter soils and lower wash temperatures of the U.S., they are decidedly noticeable under European washing conditions. One area in which U.S. detergent formulations are moving closer to those in Europe and Japan is in the use of enzymes. Because of their ability to tackle proteinaceous and carbohydrate stains on laundry, either as presoaks or formulated into laundry detergents, the materials could tackle and enzymically decompose for further cleaning such stains as blood, egg, gravy, and the like. Enzymes promptly were incorporated into detergents, and now probably two thirds of all premium-quality heavy-duty detergents contain enzymes. Japan's detergent makers reformulated all alkyl sulfonates (linear end branched) (Tables I.1–I.3).

Fatty alcohol ethoxylates and fatty alcohol sulfates with up to 10% soap in European formulations of ethanolamines, such as coco-mono-ethanolamine, contain silicones at higher levels. Some correlations use only sodium tripolyphosphate, others have mixtures of tripolyphosphate with otter phosphates (for example, sodium

TABLE I.1
Global detergent formulations

% of total formulation		Heavy-duty powders—U.S., Canada, Australia	Heavy-duty powders—South America, Middle East, Africa	Heavy-duty powders—Europe	Heavy-duty powders—Japan	Heavy-duty powders, liquids	Light-duty powders—Europe	Light-duty liquids	Detergent laundry bars
Actives		8–20%	17–32%	8–18%	19–23%	15–50%	14–25%	15–37%	18–30%
Foam boosters		0–2%		0–3%		0–4%		0–5%	
Foam depressants				0.3–5%	1–2%				
Builders									
	Sodium tripolyphosphate	25–35%	20–30%	20–35%	0–15%	10–30%	10–40%		
	Mixed non-phosphates	15–30%	25–30%	20–45%	0–20%	15–20%			25–45%
	Sodium carbonate	0–50%	0–60%						15–30%
Anti redeposition agents		0.1–0.9%	0.2–1%	0.4–1.5%		0–2%	0–0.9%		0.3–2%
Anti corrosion agents		5–10%	5–12%	5–9%	15%	2–5%	0–8%		3–8%
Optical brighteners		0.1–0.75%	0.08–0.5%	0.1–0.75%		0.1–0.5%	0–0.3%		0.05–0.3%
Bleach				15–30%					
Enzymes				0–0.75%		0–1%			
Moisture		6–12%	6–13%	4–20%		Balance	6–14%	Balance	3–14%
Fillers		20–45%	10–35%	5–45%			15–45%		

TABLE I.2
Detergent intermediates capacity far outstrips demand

Linear alkylate capacity (thousands of metric tons)				Fatty alcohols capacity (thousands of metric tons)			
U.S. producers		Western Europe producers		U.S. producers		Western Europe producers	
Conoco	170	Shell	175	Shell	210	Henkel	120
Witco	20	Chimica	140	Vista	106	Condea	60
Effective	272	Effective	615	Effective	390	Effective	305
Demand	240	Demand	325	Demand	210	Demand	160

orthophosphate), aluminosilicate zeolites, nitrilotriacetic acid, sodium citrate, and sodium carbonate; others use builders alone or in various combinations. For example, sodium carboxymethyl cellulose, other cellulose-based polymers, or synthetic polymers (e.g., sodium silicates as purchased solutions) in the U.S. or sodium perborate. In Europe, formulations for high-temperature washing have sodium perborate. Whereas those for use at lower temperatures include boosters such as tetraacetylethylenediamine. Most formulations include 20% to 25% bleach and up to 0.75% enzymes, as purchased granules predominantly contain sodium sulfate, and also include small amounts of coloring agents, opacifiers, fragrance, and viscosity builders, but can also include other filler materials such as kaolin and talc.

As the environmentalist campaigns against phosphates are rekindled in some areas in the U.S., and as phosphate limits are tightened in some European countries, the most frequently used phosphate builder is sodium tripolyphosphate. Not only it is an excellent sequestering agent, but it helps during the manufacturing process. Because it forms a stable hydrate, it allows the product to contain relatively high amounts of moisture without sticking in the equipment. Because it has been used for so long, detergent manufacturers have essentially worked its cost into the selling price of laundry powders. Its cost-efficiency becomes the standard which other products must match or better.

Zeolite-built formulations, either using zeolite as a co-builder with phosphates or as sole builder, in Europe are basically being sold only where there are phosphate limits. There would be little likelihood of finding them, for example, in the U.K., because there is no eutrophication problem there, hence no phosphate limits. Industry observers say that, that is what led P&G to reformulate the zeolite-built products it had been marketing in phosphate-limit areas. Producers of zeolites in the U.S. had high hopes for P&G's success with the zeolite formulation. If P&G made it, the reasoning went, the market would quickly expand throughout the rest of the industry. P&G's reformulation, back to phosphates—to a level of 8.2% phosphorus, just under most of the limits of 8.7% phosphorus—dashed those hopes and left some nearly new zeolite plants woefully underutilized. On the other hand, P&G is continuing its limited test-marketing of sodium nitrilotriacetate (NTA) as a builder for detergents in Indiana, where phosphates are banned. It had been test-marketing

TABLE I.3
Detergent formulations around the world have great variety

% of total	Heavy-duty powders	Heavy-duty powders	Heavy-duty powders	Heavy-duty powders	Heavy liquids	Light liquids	Detergent laundry bars
Formulation	US, Canada, AUS	Mideast, Africa	Europe	Japan			
Actives	8–20%	17–32%	**8–18%**	19–23%	15-50%	15-37%	18-30%
Foam boosters	0–2%		0–3%		0-4%	0-5%	
Foam depressants			3–5%				
Builders							
Sodium tripolyphosphate	25–35	20–30	20–45	0–15	30-40	25-40	
Mixed or non-phosphate	15–30	25–30	20–45	0–20	15-20		25-45
Sodium carbonate	0–50	0–60					15-30
Anti-redeposition agents	0.1–0.9	0.2–1	0.4–1.5		0-2	0-0.9	0.3-2.0
Anticorrosion agents	0–10	0–12	0–9	0–14	0-5	0-8	0-5
Optical brighteners	0.1–0.75	0.08–0.5	0.1–0.75		0.1-0.5	0-0.3	0.05-0.3
Bleach			15–30				
Enzymes			0–0.75		0-1		
Moisture	20	13	20		14		14
Filler	20–45	20–35	20–45	15–45	20-45	20-50	20-30
Other ingredients						20-30	15-20

detergents with NTA in New York state, another ban area, but backed off there in fear that the state would legislate a ban on NTA as well. NTA has been tested and debated as a replacement builder when the phosphate issue first arose. It is already in full use in Canada and in Finland. But because of early fears about carcinogenicity, or about heavy metals carried into the drinking water supply by superefficient sequestrant NTA, U.S. officials have never smiled on the material. The Environmental Protection Agency has said only that it sees no reason to ban the material, a statement far short of endorsement. However, the New York State Environment Commission has decided to look into the matter, and labor unions and consumer groups are still actively anti-NTA. Consumers in Europe also tend to have a similar reaction to NTA, except in Finland. Perhaps for that politically powerful reason, most governments have advised caution over wholesale introduction of the builder as a replacement or an extender for tripolyphosphate.

For the consumer, modern bleach systems are a vast improvement on the standby of "lawn bleaching"—spreading laundry out on the ground to allow the sun to do the bleaching. But the washing habits of different countries have led to the development of two distinctly different philosophies of bleaching: one of adding bleach as a separate product and the other of incorporating it into a laundry-detergent formulation. For the U.S., the industry and consumers historically have tended to regard bleach as a separate item to be added along with the detergent or used separately as a pre-laundry soak.

In Europe, on the other hand, detergent makers have incorporated bleach into their formulations for many years. The German detergent maker Henkel's premier brand Persil, in fact, was a contraction of sodium perborate and sodium silicate, indicating the existence of bleach in the formulation when it originally was developed as a soap powder. Perborate as the bleach fit in nicely with European washing habits. The cottons and linens most common then could be put "to the boil", at 95 °C, at which temperature the perborate would be more than 90% effective as an oxygen bleach for wine stains and heavy soil stains. And even if lower temperatures were used, for the small percentage of the wash load that was colored, the temperatures were still high enough that most of the perborate efficiency was retained. With that kind of formulation to depend upon, then, the consumer seldom needed special bleaching products, such as liquid chlorine. Chlorine, with its typical odor, came to be associated with disinfection, hospital cleaning, and other similar applications.

In the U.S., washing temperatures were never so high as those in Europe. More typically, "hot" was around 55 °C, and at such relatively low temperatures, perborate is only about 60% effective. Moreover, clothes and linens tended to be used for shorter periods of time before they were washed, so not all of them needed bleaching. Those factors encouraged the development of separate bleaches for use only when an actual need was perceived. Several detergent brands are on the market in the U.S. with an oxygen bleach in the formulation, but they are not really typical.

In both the U.S. and Europe, the industry has been facing some problems with bleaching technology for several different reasons. In the U.S., the rapid rise of synthetic fibers, which can be affected adversely by chlorine bleach, has given a boost to oxygen bleaches. Such products have been around for years, dry powders in the shadow of liquid chlorine bleaches. In Europe, a general drop in washing temperatures has meant a decrease in perborate bleach efficiency. Here, the answer has been the development of bleach activators, to help boost performance back to nearly the level of that "at the boil". Several boosters are being used in Europe: penta-acetyl glucose, frequently found in private-label brands, and tetra-acetyl ethylene diamine (TAED). The TAED releases into solution the unstable peracetic acid, which becomes the bleaching agent at the lower temperatures.

Oxygen bleaching in both the U.S. and in Europe has been the province of sodium perborate, and manufacturers of any other oxygen-bleach materials have had to struggle to gain acceptance. Sodium percarbonate has been slowly gaining some use in Europe, in particular, but it may be in the specialty field that has been found as winner: magnesium mono-peroxy-phthalate hexahydrate. The material, which releases the peroxyacid anion in solution, has been adopted by Henkel for its new specialty detergent for delicate fabrics and hard-surface scouring cleansers. For products for use at higher laundry temperatures, however, European detergent makers are sticking solidly with perborate and TAED.

The evolution of surfactants continues after 40 years. Synthetic detergents were introduced in areas where water hardness and mineral content were the highest, and the resulting soap curd problems were the most severe. The surfactants were anionics—first, sulfated alcohols, and, later, sulfonated alkylbenzenes. As detergents continued to spread from hard-water areas to regions with softer water and less soap curd occurrence, anionics consolidated this hold on the synthetic-detergent formulations. The anionics had strong, rich lather, a characteristic that consumers already associated with cleaning efficacy. And they were, indeed, efficient cleaners of the fabrics found in the average wash load—cottons, by and large. Cottons accounted for more than 70% of the world's production of textile fibers, with synthetic fibers at about 18% and wool 10%.

Detergent makers and intermediates suppliers had begun running into problems. First, synthetic fibers, and the rise in consumption of nonionics, started to follow that of the new fibers. Nonionics also got a boost in the U.S. by lowering the cost of nonionic. Better resistance to hard-water ions than the anionics helped them move into formulations when phosphate bans began hampering detergent efficacy. Since nonionics are easier to incorporate into liquid detergents than are anionic surfactants, they got a boost when liquid heavy-duty detergents began their dramatic rise in popularity.

Natural fatty alcohols provided the only source of raw material for making detergent-range alcohols. But petrochemical products had started to make their inroads, and now naturals are only about a third of the business. Procter & Gamble, with its captive 98,000 metric-ton-per-year plant, and Henkel, with its estimated 170,000 metric-ton-per-year plant, both use natural feed stocks. Sherex uses naturals for its alcohols, and Condea in West Germany has about half its capacity suited for making alcohols from naturals.

The synthetic detergents industry was hit hard with the problem of biodegradability. The first generation of alkylbenzenes developed for sulfonation were branched chain, hard for microorganisms to break down in sewage disposal plants and in nature. Sewage treatment plants started seeing foam buildup as a result, and foam also began appearing in rivers and streams.

There was not a serious question of toxicity, just of the biodegradability of the materials. Producers voluntarily agreed to switch their formulations to analogs of the branched-chain surfactants, linear alkylbenzene sulfonates in the U.S. Similar actions took place in Western Europe at about the same time. Anionics suppliers weathered that convulsive shift reasonably well, but the boom in synthetic fibers was another story. Synthetics had serious growth, accounting for about 38% of the world's production of textile fibers. Cotton's share had begun shrinking, hitting about 62%. Detergent makers decided that, as good as the anionics were at cleaning cottons, the nonionics were better at cleaning other fabrics.

Equilibrium to the industry as shifts a gain by nonionics at the expense of anionics was taking place. In the developing countries, there is still a strong growth, a good deal of reformulation going on, as developing countries shift their production and consumption to biodegradable surfactants. Estimates that demand for branched-chain alkylbenzenes is still increasing, mostly in Latin America and the Far East. Although there is a global pricing pattern to the surfactant industry, little product actually travels back and forth, but rather is usually consumed close to where it is produced. One exception is branched-chain alkylates, for which the augmenting natural supplies, although a variety of reasons have caused most such plans to be shelved. Naturals offer the lure of a continuing good raw material resource, renewable and manageable.

The reduction of phosphates is opening up a space in the detergent formulation that must be filled. Moreover, the percentage of surfactant in the formulation has been trimmed slightly, again opening room in the formulation. That, in turn, is putting increasing demand on producers of filler, or carrier, materials to help make up that 100% of product that goes in the detergent box. At the top of the list is sodium sulfate, a neutral, inert carrier for the active ingredients. Sodium carbonate could also be considered a filler, although it does have precipitating action against some water-hardness ions, so also functions as a builder. And in some products—laundry soap and detergent bars, for example—sodium bicarbonate is the filler of choice.

Soap bars and the increasingly popular synthetic detergent bars still are used heavily in developing countries. They provide concentrated detergents on the fabrics, with little waste, and are also easy to carry about. Slurry-derived pastes of synthetic detergent were sold in the Indonesian market. Paste was liked and accepted by the consumers in that market. Indeed, perhaps because washing and doing the laundry are such a part of home life, habits tend to change only very slowly. Regulations could, presumably, force massive changes, but fortunately for the detergent industry, the industry has weathered governmental actions through shifts in ingredients, perhaps, but not drastic changes.

The major directives affecting the detergent industry from the European Economic Community, for example, have covered the biodegradability of surfactants. The first such directive covered anionic surfactants, to ensure that all anionics used be 80% biodegradable. The next directive provided similar rules for nonionic surfactants. A third for cationics. The industry is covered by EEC regulations, and synthetics dominate world output of fatty alcohols production, thousands of metric tons. Directives on dangerous substances—what is being used in the formulation and what must therefore be put on the package label.

The industry tends to concentrate on the vast quantities and the more recent marketing excitement of the synthetic detergent developments part of the business. But soap continues to be a force that cannot be ignored. There will be an estimated 16.96 million metric tons of soaps used worldwide. Admittedly, that is down from the 17.87 million metric tons. Most of the use is in the developing countries. The detergent industry in South Korea, for example, is dominated by soaps, as are those in many African countries. Abundant supplies of natural fats—tallow for example–can lead to political endorsement of soapmaking, as in Latin America. Nevertheless, even in the U.S., about 1,500,000 metric tons of soap and soap products will be used. In the U.K., soap is the base of one of the country's leading heavy-duty laundry detergents, Unilever's Persil. Sold as a formulated soap powder, such a product is well suited to the twin-tub machines that still account for slightly more than half the washing machines found in the U.K.

I.1 HISTORY OF SYNTHETIC DETERGENTS

The need for a soap-like product which would not form curds in hard water had long been recognized, but to most chemists of the early 1920s it was probably just a tantalizing dream, Nevertheless, a definite research effort in this direction was made by P&G in the late 1920s, and from later results we now know that this work came very close to success. Coconut monoglyceride acid sulfate was prepared and was expected to show soap-like properties when neutralized to form the sodium salt. Unfortunately, the conditions used in the tricky neutralization step were not quite correct and, while the desired material must have been formed, it was immediately converted into ordinary coconut soap. Although the theory was correct, the difficulties were not appreciated at the time and the effort was considered a failure (some years later this material was actually marketed as a detergent). Thus P&G, and the industry in general, were conscious of the need for a good hard-water detergent but were not optimistic that a good product could be found.

Soap, the first detergent, has been known at least since the time of the ancient Romans. The product of the early soap-makers was a luxury item often subject to heavy taxes. It was not until the rapid expansion of centralized manufacture and skilled selling in the late 19th century that soap became a low-priced necessity in the majority of homes.

In addition to its household use, soap was used over the years in many industries as a detergent and wetting agent. The textile industry has always been the largest industrial consumer requiring detergents for washing, scouring, and de-sizing operations, and wetting agents for dye leveling, mercerizing, and bleaching. Soaps are not ideal for these purposes. Although they perform well in warm, soft water, their calcium and magnesium salts precipitate from hard water, depositing on the textile fibers and impeding the penetration of solutions. The resulting unevenness in the finishing of fabrics made the textile industry receptive to products which promised to overcome the deficiencies of soap.

Frémy observed that unsaturated oils, notably olive oil, react with concentrated sulfuric acid. Three years later, such a product was neutralized to obtain a wetting agent that greatly aided the dyeing of cotton textiles with certain red colors. These Turkey Red oils, as they became known, were the first improvements on soap for use as wetting agents, but they were not good detergents.

The next improvement came with the discovery that the sulfate esters of the fatty alcohols from spermaceti oil had a high tolerance to hard water and that their solutions would rapidly wet out fabrics. The first sperm-alcohol sulfates were prepared in 1836 but not until the end of the 19th century were they used in the textile industry as dyeing assistants, as wetting agents, and as detergents, to replace soap in special applications. They were high-cost specialty products, whose manufacture and use were of no interest to the soap makers.

Under the pressure of the fat shortage during World War I, German scientists examined the possibilities of many synthetic washing compounds. I.G. Farben investigated the practice, based on the old folklore of certain backward mountain areas in Germany and Spain, of using bile from slaughtered cattle as a wetting agent for textile dyeing. They found that the ingredient which gave this performance was a taurine derivative. A closely related product was synthesized and sold to the textile trade as Igepon T. It was a good wetting agent, a fair detergent, not affected by hard water, and resistant to acid, but it was also expensive and had physical characteristics that were unsuitable for home detergent use.

During the 1920s, other German companies developed the production of saturated fatty alcohols from fatty acids and their esters by catalytic hydrogenation. The finished sodium alkyl sulfate was sold in 1931 to the textile trade as a detergent and a finishing aid.

In 1931, P&G evaluated the German sodium alkyl sulfate for possible application in household products. License agreements were negotiated with the German patent holders, and the jointly owned Gardinol Corp. was formed by P&G and DuPont, who also held patent applications on processes for making fatty alcohols. These agreements, signed early in 1932, left P&G free to develop and market alkyl sulfates as synthetic detergents in the household and laundry fields.

Dreft, introduced in 1933, was the first American synthetic detergent packaged for household use. It was a free-flowing powder consisting of about 20% sodium coconut sulfate with sodium sulfate and water as the rest of the formulation. Drene, a liquid shampoo, was introduced in 1930. Both Dreft and Drene found market acceptance but the volume of sales was limited. Although Dreft and other light-duty synthetic detergents did not form hard-water curds and cleaned lightly soiled garments adequately, they would not properly clean medium or heavily soiled laundry. A minor gain in detergency could be made by adding builders such as those used to improve the detergency of soap, but they left the clothes harsh and stiff. The builders reacted with the hardness in the water to form insoluble granular materials which adhered to the clothes. When used as builders with soap, it is mostly the soap that reacts with the hardness to soften the water. The scum formed is objectionable and adheres to the clothes but the resultant "feel" is not so objectionable. A suitable builder for synthetic detergents was finally found in sodium tripolyphosphate. Replacement of a large part of the sodium sulfate by tripolyphosphate built the product to the point where detergency approaching that of soap was attained. In the presence of the polyphosphates, smaller amounts of organic surfactant could be utilized with an improvement in mildness and without sacrifice of sudsing or soil-removal properties.

Several problems had to be resolved before a heavy-duty detergent could be marketed. At the time of its development by P&G, tripolyphosphate was only a laboratory chemical and was not commercially available. Even if the phosphate were available, new plant units had to be built to process the new product. Consequently, it was not until after World War II that the building of both phosphate plants and new spray-drying towers was started.

Test marketing of Tide, the first all-purpose laundry synthetic detergent, was started in the sixties. Two shortcomings were noticed. One of these was that soil removed from the laundry was not held in suspension properly, a portion redepositing on the fabric to give the laundry a gray cast. Prevention of soil redeposition was accomplished by the inclusion of sodium carboxymethyl cellulose in the formulation. Another problem was that corrosion of certain metals was stimulated by the capacity of Tide to solubilize and remove the metallic oxide films.

These films form on metals and give them a degree of protection against further corrosion. Sodium silicate was added to prevent aluminum corrosion (e.g., in automatic washers) and benzotriazole to prevent the tarnishing of tableware.

The introduction of Tide initiated the dramatic replacement of soap as the general household laundry detergent. It also stimulated much activity in finding useful substitutes for soap, not only by other soap makers but by chemical and petrochemical companies as well. For example, coconut fatty alcohol was replaced by tallow fatty alcohol to give a quality improvement at a reduction in cost. Also, tetra propyl benzene from petroleum sources was made available in large quantities, and sodium alkylbenzene sulfonate has now become the most widely used of all surfactants. It has a relatively low cost and, since it is a petroleum-based product, its supply is assured and the cost is more stable than that of fats.

One drawback of most of the first heavy-duty detergents was that they produced an over-abundance of suds when used in front-loading automatic washers. New formulations containing marine fatty acids, as suds depressors, and non-ionic surfactants, as detergency builders, solved this problem. Dash, introduced by P&G in the 50s, is of this type.

Further demands for convenience led to the development of a pressed tablet of granular detergent which gave the consumer a pre-measured quantity of product. Salvo, a low sudsing tablet having a Dash-type formula, was introduced in 1959 and expanded nationally in the U.S. in 1962.

The successful development of detergents, resistant to hard water, has been very profitable not only for P&G but for the many other companies that have participated in the opening up of a new field of chemistry. New products continually are being made and tested.

Those who have gained the most, collectively, are the millions of consumers who use and enjoy these new and improved products.

I.2 DEVELOPMENT OF SYNTHETIC DETERGENTS IN THE INTERNATIONAL MARKETS

The international markets were created with responsibility for the administration and operation of the manufacturing, marketing, and advertising functions in all areas outside the U.S. These new markets have grown rapidly with a potential for still greater expansion. Synthetic detergents joined the list of international products soon after their introduction in the United States.

1.3 INTERNATIONAL FACTORIES PRODUCING SYNTHETIC DETERGENTS

Production by subsidiaries is complemented by contract manufacturing operations in many developing countries. In addition, several packaging operations and a vigorous export trade administered by subsidiaries round out the international synthetic detergent business.

During the early stages of international operations, the research and development work was carried out by U.S. domestic organizations. As the international business began to grow, the problems involved became more complex. Available raw materials, marketing conditions, and washing conditions in various countries all differ sufficiently from each other to demand special formulations processing or packaging. Thus, while initially, formulations, processing, etc., of the international brands were very similar to those in the U.S., numerous modifications unique to the international markets have since been developed.

The author has tried to incorporate as many new concepts and developments as possible while at the same time trying to keep the size of this publication within reason and make it easier to read and understand. The experienced reader will notice immediately that many of the graphs and drawings have been incorporated within the written text.

Sulfonation has been discussed to recognize the move to higher oleum strengths and to the use of sulfur trioxide as a sulfonating agent. High active formulations have been changed to acid mix concentration since the use of higher strength sulfonating agents and sodium sulfate removal is covered. Paste and crutcher mix phase systems have been discussed in order to present a more basic discussion of the surfactant water-electrolyte phase relationships. "Crutching" and "spray drying" are included with information on the role of phosphate hydration and more information on the procedures necessary for blowing lighter density granules and low sudsing products. Processing Steps on perborates include the importance of cold-mixing. Finished product quality and trouble-shooting checklists are added. Fundamental principles of product caking and predicting product behavior is discussed.

1 Detergent Raw Materials

Synthetic detergents are formulated to give the most desirable product and performance properties consistent with cost and availability of raw materials. Each of the many ingredients performs one or more specific functions. Many of the ingredients listed are commonly termed "minor ingredients". This only means that their weight usage is relatively low. Their effect on performance and, therefore, on consumer acceptability is in no sense minor but may be far more important than other constituents (e.g., sodium sulfate) present in much larger amounts.

Synthetic Detergent Ingredients and Their Functions

Ingredient	**Function**
Surfactant	Detergency, sudsing
Inorganic builder	Detergency
Suds depressor	Reduce suds volume and stability
Silicate solids	Prevent aluminum corrosion. Improve physical characteristics of product
Sodium toluene sulfonate	Improve processing. Anti-caking agent
Sodium perborate	Bleach
Magnesium sulfate	Anti-caking agent for humidity caking
Carboxy methyl cellulose	Detergency (prevent soil redeposition)
Brightener	Improve whiteness of washed fabrics
Tarnish inhibitor	Prevent tarnishing of silver, brass
Pigment	Increase consumer acceptability of product. Call attention to a product improvement
Dye	Improve whiteness of washed fabrics
Perfume	Improve consumer acceptability of product Cover the base odor of the product
Bacteriostat	Prevent odor development on fabric or in wash water
Sodium sulfate	Diluent brought in during processing Cheapen formula
Borax, sodium carbonate	Cheapen formula (carbonate reduce dust)
Water	Cheapen formula Prevent dustiness
Sodium aluminosilicate	Anti-caking agent

DOI: 10.1201/9781003596219-2

1.1 SURFACTANTS

The surfactant is the basic cleaning agent in a synthetic detergent formulation. An important property is its degree of sudsing during use since the consumer often relates sudsing to cleaning.

1.1.1 Classification of Surfactants

The organic active materials used in synthetic detergents are surface active agents (surfactants) with a high cleaning ability. Chemically, they contain a long, water-insoluble (hydrophobic) chain, usually hydrocarbon, with a water-soluble (hydrophilic) group at one end (Table 1.2). Surfactants are commonly classified as follows.

1.1.1.1 Ionic

Compounds of this type ionize in water. They are subdivided into three classes:

Anionic: The active portion of the molecule, containing the hydrophobic chain and hydrophilic group, is negatively charged in solution.

(alkylbenzene sulfonate) 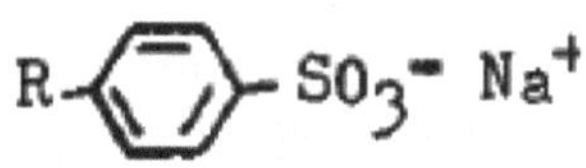

(alkyl sulfate) $R-O-SO_3^- \quad Na^+$

(Soap) $R-COO^- \quad Na^+$

In general, the detergency and sudsing properties of the anionics are superior to those of other types.

Most commercial granular detergents are made with this type; however, many structural varieties exist. Soap, of course, is a member of this class.

1.1.1.2 Cationic

The active portion of the molecule, containing the hydrophobic chain and hydrophilic group, is positively charged in solution.

$$R\,(CH_3)_3\;N^+\;Cl^-$$

Example:
These compounds are used primarily in detergents for germicidal and disinfectant applications. Cationics are poorer in detergency, less mild, and usually more expensive than anionics.

Ampholytic (Amphoteric): The active portion of the molecule can be cationic or anionic, depending on the pH of its solution.

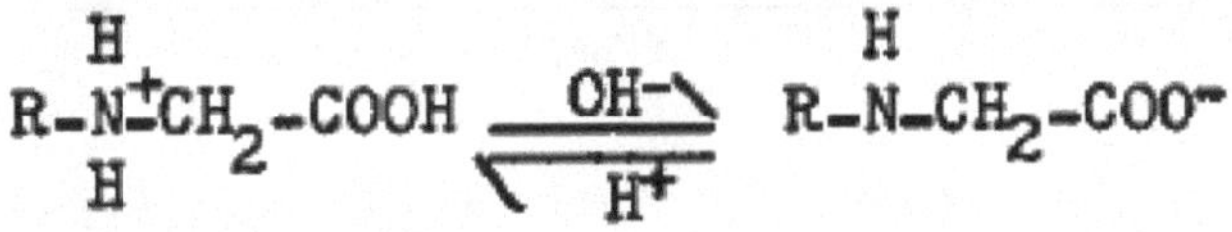

Example:

R–N–CH–COOH   R–N–C–COOH

At present, only a few compounds of this type are available and their performance characteristics are still largely unknown.

1.1.1.3 Nonionics

These compounds do not ionize in water. The most widely used materials of this class are made by addition of one or more ethylene oxide and/or propylene oxide groups to the hydroxyl group of alkyl phenols or aliphatic alcohols. The ether group, resulting from the addition, forms the hydrophilic part of the molecule. Many varieties of these compounds are available, but they are generally less effective in cotton detergency than anionics and generally low sudsing. Their primary use in detergent granules is as a suds depressor for low sudsing products.

1.2 PROPERTIES OF SURFACTANTS

Properties of surface active agents often are difficult to evaluate because of the interrelation of their characteristics.

Detergency generally improves with increasing hydrophobic chain length. In most cases, however, there is a point where the solubility decreases faster because of the higher molecular weight, resulting in a net decrease of the detergency. Sudsing is often regarded as indicative of cleaning power.

In automatic washing machines with rather violent agitation, excess suds can interfere with the cleaning action. Hence, the quantity of suds, speed of sudsing, and suds stability of the product are varied to suit the intended use. Sudsing depends on the particular active used, its chain length, the amount of branching in the chain, and the washing conditions (water temperature and hardness). Solubility decreases with increasing molecular weight, but other generalizations cannot be made.

Mildness depends primarily on the type of active. It improves slightly with increasing molecular weight. Although all surfactants used in consumer products range from only slightly irritating to very mild, there is some difference. This is expressed below on an arbitrary 1–10 scale in which 10 means no reaction.

TABLE 1.1
Problems in Raw Material Quality

Material	Defect	Effect on Process or Product	Comment
Oleum	Low strength	Low sulfonation completeness	Use more acid if formula permits
	High strength	Burning in sulfonation, poor color and odor%	Reduce usage, add NaS0 in crutcher
	High iron content	Poor product color and odor	
Alkyl benzene	High bromine number	Poor product color and odor due to reaction at the points of unsaturation	
	Wide boiling range	Heavy fraction: poor sulfonation completeness. Poor sudsing in cool water washing. Light fraction: poor detergency	
Water	Present	Low completeness due to oleum dilution	Settle and drain off water
	Incorrect specific gravity	Indicates contamination by other hydro-carbons. Could result in off-quality product	
	Impurities	Product contamination	
	Positive Doctor test	Indicates presence of sulfides. Poor color and odor	
Fatty alcohol	Water present	Dilutes oleum, causes low completeness	Settle, but water is soluble to about 1%. Use more acid if formula permits
	Incorrect hydroxyl value	Incorrect molecular weight may affect sudsing and/ or detergency	
	Visual impurities or incorrect specific gravity	Indicates contamination. Could affect product quality	

TABLE 1.1 (Continued)
Problems in Raw Material Quality

Material	Defect	Effect on Process or Product	Comment
Toluene	Incorrect density or boiling range	Indicates contamination. Can cause off-quality product	
	High acid wash color	Poor finished product color	
	Positive Doctor test	Indicates presence of sulfides. Poor color and odor	
	Impurities	Product contamination	
Caustic soda	Low NaOH	Economy. Paste moisture too high	Reduce paste water usage
	High sodium carbonate	Excess CO_2 in paste causes spray drying problems	
	High Fe, Cu, or Mn	Poor color and odor	
	High NaCl	Diluent. Process equipment corrosion	
Sodium carbonate	Low Na0H content	Economy	
	High NaCl	Dilutes product	Reduce acid ratio in paste or crutcher mix
	High iron or other impurities	Poor product color. Difficulties in solution handling	
Sodium tripolyphosphate	Low P_2O_5	Economy. Product composition	Increase usage and adjust sulfate
	Low STPP	Reduces detergency (phosphates present in less effective form)	
	High temperature rise	Can cause lumping or grit in crutcher	Increase paste temperature
	Low temperature rise	Insufficient hydration in crutcher	Use colder paste, allow more aging time
	Positive sulfide or wet odor test	Poor product odor	

(continued)

TABLE 1.1 (Continued)
Problems in Raw Material Quality

Material	Defect	Effect on Process or Product	Comment
	Coarse material	Poor hydration, processing difficulties, grit in the finished product	Pulverize or screen
	Miscellaneous impurities	Poor product quality	
	Low pH	Low product pH	
	High loss on ignition	Impurities in phosphate. Reduces temperature rise test. Can cause lumping and other processing difficulties	
Sodium silicate	Low solids content	Economy. Poor control of product composition	Increase usage
	High ratio	Excessive floc	
	Low ratio	Soft or sticky granules. Possibly inadequate protection against corrosion of aluminum. Might cause high product pH	
	Miscellaneous impurities	Harms product quality, notably color	
Sodium sulfate	Low Na-sulfate	May affect processing	
	High NaCl or Na-carbonate	Poor quality product. Can affect spray drying	
	High moisture (or incorrect moisture)	Economy	Adjust usage
	Iron and other contaminants	Harms product quality	
Sodium perborate	Low available oxygen	Economy. Poor product performance	Increase usage. Adjust sulfate
	Out of sieve test limits	Harm product appearance and grit	
	Caked or lumpy	Handling and metering difficulty	Break up and screen
Amide	Low amide content	Economy. Poor product performance	Increase usage and adjust sulfate

TABLE 1.1 (Continued)
Problems in Raw Material Quality

Material	Defect	Effect on Process or Product	Comment
	Poor color	Can affect product color	
	High free fatty acid	Depress suds	
CMC, benzotriazole, brighteners	Low activity	Poor performance in product	Increase usage
	Caked, lumpy	Processing difficulty	Break up, screen
	Visually different from standard	Possibly poor product performance	
Perfumes	Odor different from standard	Poor odor in finished product	Can possibly be blended

Notes:
The comment lists the corrective action that may be taken by the factory after due authorization when it is decided not to reject a raw material which does not meet specifications.
Economy denotes that the defect is merely one of reduced active level due to inert ingredients.

Pure actives:
Coconut alkyl sulfate
Coconut sodium soap
Alkyl benzene sulfonate
Coconut alkyl glycerol ether
Tallow alkyl sulfate / sulfonate
Monoglyceride sulfate

Additional properties of surfactants, such as stickiness, stability under storage and use conditions, and biodegradability all are important and must be carefully considered when formulating a new product.

1.3 SURFACTANT RAW MATERIALS

The two types of actives most often used in detergent granules are alkyl benzene sulfonates (ABS) and alkyl sulfates (AS).
The materials required for their manufacture are:

(a) Alkyl benzene, which is sulfonated to make ABS, is a hydrocarbon of the form:

The chain R:

(i) Normally averages about 12 carbon atoms but can average up to 15 carbon atoms,
(ii) Usually consists of a mixture of chain lengths varying from 9 to 15 carbon atoms,
(iii) Shows a varying and often, as in the tetrapropylene ABs, a high degree of branching.

At present, there are three major types of AB on the market:

Regular tetrapropylene AB, averaging 12 carbon atoms in the alkyl chain, is the most widely used AB. Polypropylene AB average 13–15 carbon atoms. The alkyl chain have somewhat better detergency but are slightly lower in sudsing. Straight-chain ABs are more biodegradable and are somewhat better in detergency than tetrapropylene AB but are generally slightly poorer in sudsing.

To manufacture polypropylene benzene, propylene is catalytically polymerized and the reaction mixture is fractionated to obtain the desired average molecular weight. This fraction is then reacted with benzene in the presence of an alkylation catalyst such as AlCl or HF to make alkyl benzene. The AB is then purified by distillation and usually contains at least 99% alkyl benzene; the impurities are mainly aliphatic hydrocarbons. The alkyl chain of these ABs is highly branched and the benzene ring is attached to any secondary or tertiary carbon atom.

Because both reactions, polymerization and alkylation, are done under conditions that lead to isomerization, the resulting AB is a very complex mixture with side chains varying both in number of carbon atoms from C8 to C15 and in degree of branching. The exact mixture that results depends on the conditions of both polymerization and alkylation. This accounts for the differences in AB between suppliers and to some degree even among batches from the supplier. Presumably this is the reason that ABs from different suppliers can be significantly different in performance characteristics.

The only analysis that shows these differences clearly, without identifying the exact nature of the differences, is vapor-phase chromatography (VP). It is possible to identify an unknown AB as to type and supplier by running a VPC analysis and comparing the results with a set of standard curves prepared from ABs of which type and supplier are known. It has not yet been possible to translate the differences in the VPC curves into a prediction of the performance of the AB.

Straight-chain ABs are becoming more important because they are more easily degraded by bacteria and, therefore, are more easily handled in sewage disposal systems, thus eliminating the problem of foam on surface waters and even in drinking water. The ABs are made by alkylation of benzene with a straight chain olefin by the same process as used for polypropylene benzene. The olefins are obtained by cracking crude petroleum wax or by a number of synthetic processes. Although these ABs do not have any side-chain branching, there is the same range in number of carbon atoms, and the benzene ring again can be located anywhere along the chain (except the 1 position), giving a possible 20–30 different isomers, all of which are usually present. In the case of straight-chain ABs, it is possible to determine the amount and type of each isomer present by VPC analysis, and good progress is being made toward predicting performance from the composition as determined by VPC. Performance of these ABs is similar to the tetrapropylene ABs with probably a slight advantage in cleaning and a small drop in sudsing for the straight chain.

(b) Fatty alcohol

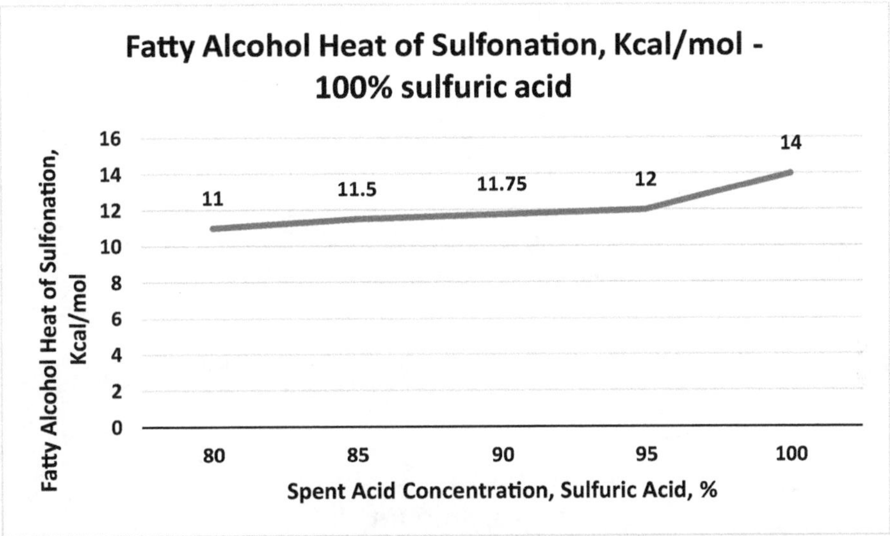

FIGURE 1.1 Fatty Alcohol Heat of Sulfonation, Kcal/mol—100% Sulfuric Acid.

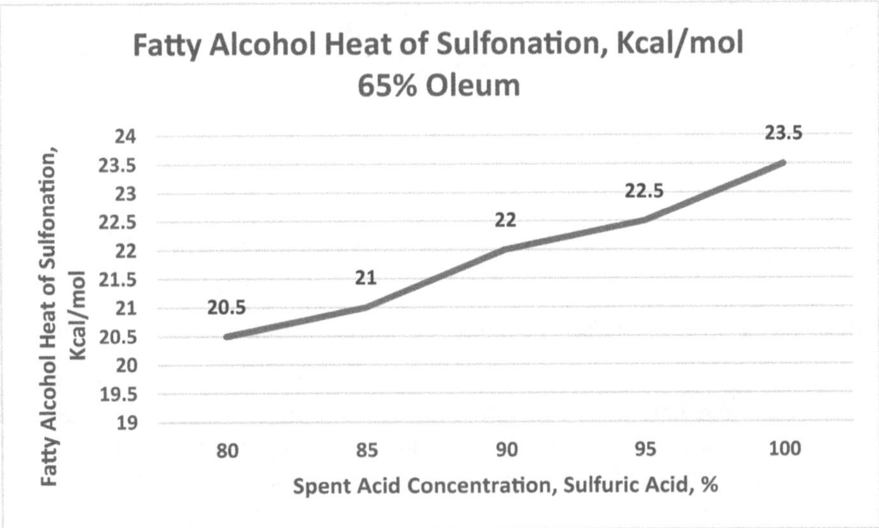

FIGURE 1.2 Fatty Alcohol Heat of Sulfonation, Kcal/mol—65% Oleum.

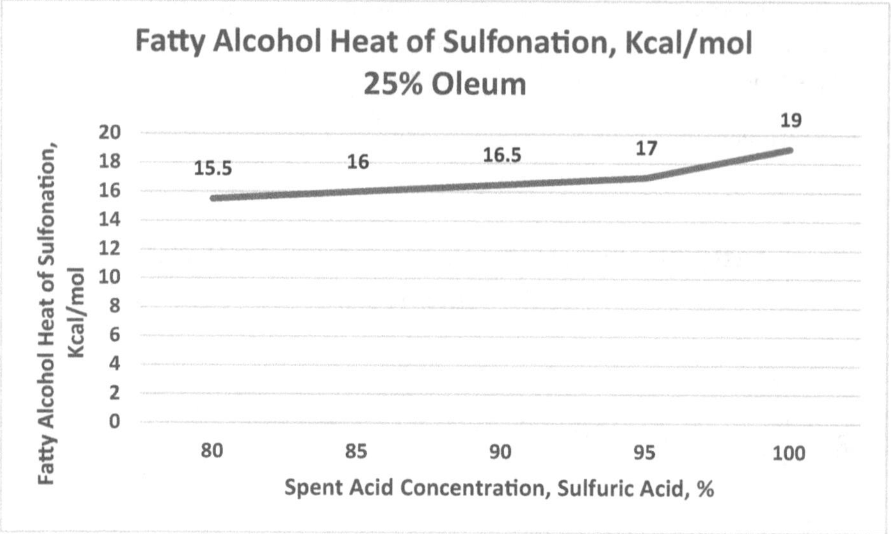

FIGURE 1.3 Fatty Alcohol Heat of Sulfonation, Kcal/mol—25% Oleum.

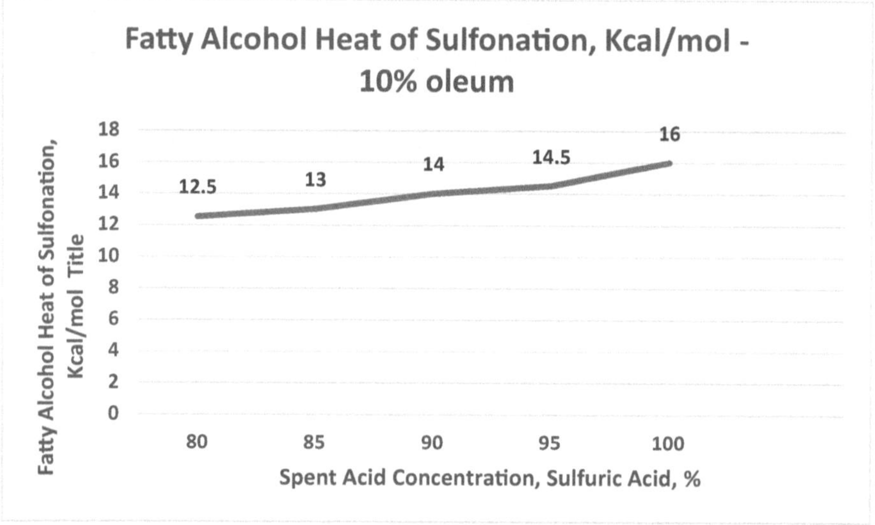

FIGURE 1.4 Fatty Alcohol Heat of Sulfonation, Kcal/mol—10% Oleum.

Both coconut and tallow alcohols are sulfated to make alkyl sulfates. These are straight chain, essentially saturated, primary alcohols of varying chain lengths.

The fatty alcohols can be made by one of three processes:

(i) Coconut oil or tallow is treated with methyl-isobutyl carbinol (MIC) to yield the fatty acid esters of MIC. These esters are then reduced with sodium to the alcohols corresponding to the fatty acids (Bouveault LeBlanc process). This method is no longer used.

(ii) The fatty triglycerides are hydrolyzed with steam under high pressure and the fatty acids obtained are reduced by high pressure catalytic hydrogenation.

(iii) The methyl esters of the fatty acids are prepared by methanolysis of the fat, followed by direct catalytic high-pressure hydrogenation of the methyl esters to the corresponding fatty alcohols.

The fatty alcohols obtained are an only even-numbered chain lengths. The weight can be calculated from the following equation:

Mixture of difference of the average molecular hydroxyl value divided by the molecular weight

The impurities in fatty alcohols vary with the process of manufacture. Both hydrogenation processes result in small amounts of hydrocarbons and some fatty acid-fatty alcohol esters. Fatty acid hydrogenation leaves some soap in the undistilled product while methyl ester hydrogenation leaves some unreduced methyl ester in the finished alcohol.

The alcohol that is sometimes used as a spray-on suds booster on alkyl sulfate products is a narrow distillation cut of coconut alcohol containing about 95% lauryl alcohol and 5% other alcohols.

(c) Oleum, or "fuming sulfuric acid", is the major source of S0 for sulfonation and sulfation. Its strength is usually expressed in one of three ways as shown below for 25% oleum.

Any strength acid from 98% to 30% oleum can be used for sulfation and sulfonation (Figures 1.1–1.4). If acid strength is much below 98%, the reaction rate is usually too slow, and above 36% oleum charring of the AB or fatty alcohol may occur.

Sulfur trioxide vapor can also be used as a sulfonating agent if it is first made less reactive by diluting with an inert gas or by carrying out the reaction under vacuum.

The strength of the acid used at each factory depends on the quality of the locally available oleum, but usually it is economically preferable to use the strongest acid (up to 30% oleum) available. Some plants prefer not to handle 30% oleum because of its high (20 °C) melting point. Depending on local conditions, this may require special heating and unloading facilities (25% oleum melts at 8 °C). Whatever the strength selected, however, it is important that recipes are uniform in strength since any variations will be reflected as changes in completeness and product composition.

Usually only trace impurities are present in oleum, but these, such as iron and nitric acid, are undesirable because of their adverse effect on color and odor.

(d) Sodium hydroxide (caustic soda) and sodium carbonate (soda ash) are used for neutralization and pH adjustment. They are used on an Na0 basis. Impurities generally are sodium salts such as sodium chloride and sodium sulfate and traces of heavy metal ions such as iron, copper, etc. The sodium salts only act as diluents and do not generally harm finished product quality; the trace metal contaminants can cause poor color and odor.

1.4 INORGANIC BUILDER (PHOSPHATE)

1.4.1 Properties of Phosphate Builders

The inorganic builder (phosphate) in a synthetic detergent serves as a detergency builder primarily in the washing of cotton goods. Several sodium phosphates are available including: sodium orthophosphate, sodium pyrophosphate, sodium tripolyphosphate, and amorphous metaphosphate glasses.

An exception is trimetaphosphate ($NaPO_4$) which is a ring, crystalline (non-amorphous) compound with no detergency building properties. In most cases, the potassium salts are also available.

The major function of the phosphate is to sequester the water hardness. However, it also provides alkalinity reserve and increases the ionic strength, both of which improve detergency.

Although the Ca and Mg ions present in hard water do not precipitate anionic synthetic surfactants (as they do soap), they do interfere with their cleaning ability. The phosphates sequester the water hardness thus freeing the surfactant. The sequestering ability of phosphates increases from the ortho to pyro- to tripolyphosphate. The amorphous glasses are equal to or perhaps better than tripolyphosphate in this respect; however, they are significantly less stable reverting to lower phosphates. For best cleaning, enough phosphate must be used to sequester all the water hardness. For optimum performance and physical properties in spray-dried detergents, tripolyphosphate is generally the best phosphate. For certain performance and use conditions, however, pyrophosphates are used in spray-dried products. Even tripolyphosphate is not stable enough for the shelf life of liquid detergents containing water. Such products usually contain pyrophosphate. The potassium salt is used because of its higher solubility.

1.4.2 Preparation of Tripolyphosphate

Sodium tripolyphosphate is a crystalline compound made by calcining a mixture of orthophosphates (Figure 1.5).

Side reactions include:

$$2NaHPO_4 = NaPO_4 + H_2O$$

$$3NaHPO_4 - (NaPO_3)_3 + 3HPO$$

Commercial STPP ranges from 80% to 95% pure. The impurities are largely the ortho-, tri-meta-, and pyrophosphates as shown in the equations above. Since the ortho- and amorphous phosphates are not desirable, the calcining conditions selected are generally such as to result in a fairly high pyrophosphate content while minimizing the ortho- and amorphous phosphate contents. In addition, the STPP may contain:

- small percentages of sulfate and chloride from the starting orthophosphates

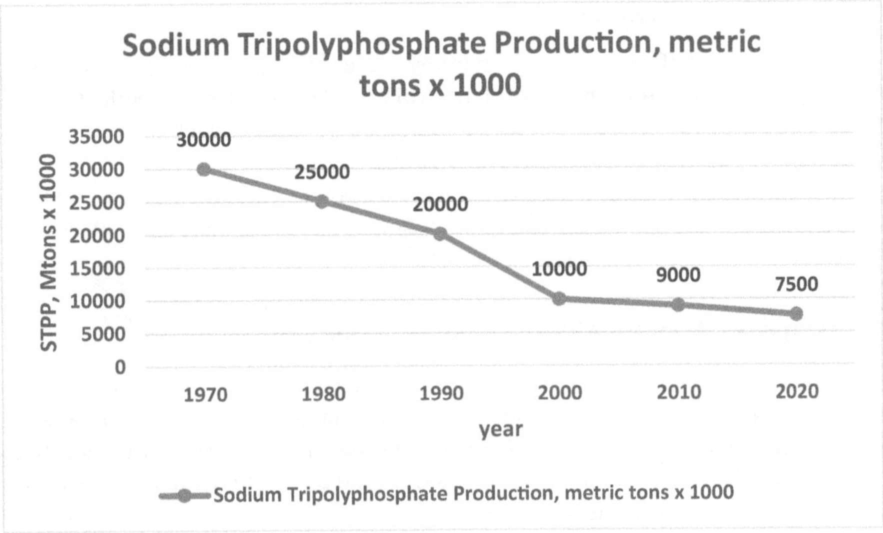

FIGURE 1.5 Global Detergent Phosphate Production.

- sulfides and phosphides which react to produce off-odors in the finished product even when present in only trace amounts and must, therefore, be minimized.

Anhydrous tripolyphosphate may exist in two allotropic forms, designated as STPP.I and STPP.II. These are chemically identical but have different crystal structures and physical behavior. STPP.II is the stable form at temperatures below about 48 °C, whereas STPP.I is formed above this temperature. Commercial

STPP is a mixture of the two forms. Since the rate of reaction from STPP.I to STPP.II is essentially zero at room temperature, both forms can exist in commercial receipts.

STPP reacts with water (vapor or liquid) to yield the hexahydrate. The hydration of STPP.I is very fast while STPP.II hydrates more slowly, but the same hydrate is formed by both. A mixture of STPP.I and STPP .II is used to control hydration in the crutcher. The maximum proportion of STPP.I is limited since too rapid hydration may form grit which will carry through to the finished product. The normal tripolyphosphates used range from 0% to 40% STPP-I.

A Temperature Rise Test is used to determine the Form-I content in a STPP sample. It is based on the fact that the hydration of STPP is a highly exothermic reaction. Conditions are chosen that inhibit the hydration of the STPP.II and, therefore, the temperature rise is proportional to the Form-I content.

For processing reasons, the LOI (loss on ignition, i.e., moisture content) of STPP must be kept very low. STPP is hygroscopic and will absorb moisture if exposed to the atmosphere. All non-bulk STPP shipments are, therefore, made in polyethylene or asphalt layer protected bags. In bulk handling systems, special precautions are taken to minimize moisture pick-up.

1.4.3 Use of Pyrophosphate

Tripolyphosphate buffers washing solutions at a high pH (9–10). Mixtures of tetrasodium and disodium pyrophosphates (pH = 12.0) are used as builders in so-called "neutral formulations". These are formulations whose water solutions have a pH close to 7. Such products are generally termed "light duty" since they do not have quite the cleaning power of products built with tripolyphosphate. They are intended for handwashing applications where mildness is important. The pH is adjusted by varying the proportions of the two pyrophosphates.

1.5 SUDS BUILDERS AND SUDS DEPRESSORS

Rapid sudsing and good suds stability are often preferred by consumers. Proper use of organic builders can at the same time make sudsing an indicator of available washing power. Different organic builders are more or less specific for various different actives. Various ethanol and ammonia amides of coconut and coconut fatty acids have proven satisfactory for alkyl benzene sulfonate.

Certain builders tend to darken on aging, especially at elevated temperatures on contact with air. Darkening occurs, for example, if coconut ethanol amide is overheated during melting. Thus, careless handling of amide can result in poor finished product color. Ammonia amide must also be kept dry since it hydrolyzes in the presence of water when hot.

Small amounts of high molecular weight fatty acids are effective in reducing sudsing for both alkyl benzene sulfonate and alkyl sulfates. They are still more effective with anionic-nonionic formulas, with which they give extremely low sudsing. These materials are termed "suds depressors".

1.6 SILICATE SOLIDS

Sodium silicate is added to synthetic detergents to provide protection against the corrosive action of alkaline phosphates on aluminum. They also increase granule firmness and crispness and improve density control during spray drying.

Silicate ion must be present in the washing solution above a critical minimum concentration to prevent the corrosion of aluminum. The active ingredient in this respect is Si0, but NaS0 is required to make the Si0 soluble. Sodium silicate is not a simple chemical entity, the weight ratio of Si0 to NaP0, commonly known as the "silicate ratio", being continuously variable. The ratio may be adjusted by mixing silicates or by adding caustic soda, but silicate solutions reach equilibrium very slowly (e.g., at least 2 hours to adjust 2.0 ratio silicate to 1.6 ratio).

Sodium silicate is incompatible with more acidic materials because the latter neutralize NaP0, causing an undesirable, flocculent precipitate of Si0 ("floc"). The higher the silicate ratio, the greater is the tendency for floc. On the other hand, increasing the silicate ratio increases granule crispness and improves processing. Hence, the choice of the correct silicate ratio and level will vary between products, depending on their requirements. Sodium silicate usually is provided as a concentrated water solution containing about 5% solids.

1.7 TOLUENE OR SODIUM TOLUENE SULFONATE

Toluene is sulfonated and neutralized to make sodium toluene sulfonate (STS). STS does not contribute to the washing process but is an important aid in processing as a result of its ability to reduce the viscosity of pastes and crutcher mixes containing alkyl benzene sulfonate. STS is of particular value with high active (35% ABS) formulas, but has no effect on, and is not normally used with, other surfactants such as alkyl sulfates. Toluene sulfonate also improves the crispness of granules and acts as an anti-caking agent in ABS formulas containing less than about 30% active.

Toluene may be considered to be a short chain alkyl benzene. Potentially, benzene and xylene are possible substitutes for toluene:

Alkyl benzene
Toluene
Benzene
Xylene

1.8 SODIUM PERBORATE

Sodium perborate tetrahydrate ($NaB0_3$ $4H_20$) is used as a bleach additive in synthetic detergent granules. It is a white, free flowing, crystalline solid containing about 10% active oxygen by weight. The active part of the molecule is hydrogen peroxide which is present as "water of hydration". The bleaching action is effective only if the wash water is at or near the boiling point. Perborate is widely used where washing is done in very hot water or where laundry is actually boiled in detergent solution.

Sodium perborates can vary considerably in stability to decomposition. Decomposition results not only in loss of bleaching power but also causes caking of perborate receipts and of the finished product. Thus, it is essential to purchase the most stable perborate possible and to minimize its decomposition during storage and processing.

1.9 MAGNESIUM SULFATE

Synthetic granules tend to cake in humid climates, and so in such climates they must be packed in expensive, protective (moisture proof) cartons. Magnesium sulfate solution, sprayed on the surface of the granules, will prevent this humidity caking in all areas where product is stored under conditions where the relative humidity averages less than about 80%. (It should be noted that conditions in storage areas such as warehouses are generally considerably less severe than official weather data would indicate.) In practice, this means that magnesium sulfate will provide anticaking protection in most hot, humid climates but not in cold, humid climates. In cold, humid climates the humidity is actually much higher (90–98%) than in the warm, humid climates (78–85%).

The value of magnesium sulfate is primarily economic in that, in areas where it is effective, the use of Mg SO with plain cartons represents significant savings over the use of the protective cartons required in the absence of Mg SO.

1.10 CARBOXY METHYL CELLULOSE

Soil removed from a fabric during washing tends to redeposit on the fabric. Carboxy methyl cellulose (CMC) aids in suspension of soil and minimizes soil redeposition. The mechanisms involved are not fully understood; however, the tendency to redeposition depends on the type and probably the chain length of the surfactant. The commonly used surfactants are listed below in order of increasing tendency to redeposition and, thus, of increasing need for CMC. The list is also in order of increasing whiteness maintenance, that is, decreasing tendency to deposit calcium salts on fabrics.

Organic Ca Salt
Tallow soap
Tallow alkyl sulfate
Coconut soap
Coconut alkyl sulfate
Alkyl benzene sulfonate

1.11 BRIGHTENERS

Fabric brighteners are complex organic materials which adsorb the ultraviolet component of daylight and re-emit it as visible light of a pinkish or bluish hue. This neutralizes the natural yellowness of aged fabrics to give an impression of greater whiteness to the cloth. Brighteners are more effective than bluing. The total amount of light reflected from a fabric with adsorbed brightener is greater, in addition to being whiter, than that from a fabric without brightener. Bluing, on the other hand, improves whiteness by absorbing part of the yellow component normally reflected from a fabric. The result is improved whiteness but reduced total reflectance as compared to a sample without bluing.

Brighteners may vary considerably in a number of important characteristics, including:

1. substantivity to (ability to adsorb on) various fibers (cotton, nylon, wool, etc.)
2. degree of fabric whitening at given usage levels
3. degree of fabric whitening under various washing conditions (time, temperature, etc.)
4. effect on granules color
5. stability to bleach (particularly chlorine bleaches)
6. stability to sunlight
7. hue imparted to cloth
8. cost

Often two brighteners are used in combination. A good cotton brightener may be used in combination with a good nylon brightener. Since the characteristics of various brighteners can differ greatly, the selection of brighteners or combinations of brighteners will vary between products, depending on their requirements.

1.12 TARNISH INHIBITOR

Sodium tripolyphosphate tarnishes metals such as brass and German silver (German silver, an alloy of copper, zinc, and nickel, is the base metal for silverplating. When the silverplate wears away, the German silver is exposed and may tarnish). The tarnish inhibitor used most commonly at present is benzotriazole (BTA).

Sodium perborate also acts as a tarnish inhibitor so that benzotriazole is not used in products containing perborate. Melamine, another inhibitor, is less effective than benzotriazole and is used in many competitive products.

1.13 PERFUME

Perfume increases consumer acceptability of a product. It masks or modifies unpleasant odors (e.g., base odor of the product), imparting instead a pleasing odor which has high acceptability yet fits product character.

Perfumes are usually blends of many materials which are relatively reactive and unstable and, therefore, have to be handled with care avoiding, above all, exposure to temperatures above 43 °C.

1.14 PIGMENT

Pigments are used for product differentiation and as "evidence". Of quality that is a major part of brand promotion (e.g., Blue Cheer whitening ability, Pink Dreft mildness, etc.). Pigments color the granules and the wash solution but are non-substantive to cloth and, thus, do not affect the washing process. Common pigments used include phthalocyanine blue and phthalocyanine green (Monastral blue and green respectively). Pigments are available in either powder or paste form. Paste is preferred for ease of handling and dispersing.

1.15 DYE

A dye is a coloring material which is substantive to fabric. For example, blue dyes (such as Polar Brilliant Blue) are bluing agents which actually color fabrics blue. Although, in this case, they enhance the action of the brighteners they are used primarily for brand promotion reasons.

In some cases, dyes are used for coloring product (e.g., Rhodamine BM for pink granules). This is done only when no suitable pigment is available. Concentrations are such that no visible coloring of fabrics is noted unless a dye touches the cloth. Very good dispersion of the dye in the product is necessary to prevent dye specks.

1.16 BACTERIOSTATS

Bacteriostats inhibit the growth of bacteria. They are not widely used in detergent formulations as yet but they have several potential applications.

1. In industrial products, bacteriostats allow detergents to sanitize as they clean. For example, in hospitals they help prevent the transmittal, through bed linen, etc., of resistant strains of bacteria from one patient to another.
2. Many cases of odor development (e.g., axillary odor, diaper odor, even stale or sour odors in stored towels, etc.) are due primarily to bacterial attack. If a bacteriostat is deposited on the fabric during washing, the residual odor and also odor development during storage and use are considerably reduced.
3. In some areas (e.g., Venezuela) where laundry is soaked overnight in detergent solution before washing, a foul odor often develops. This is due to bacterial attack of certain ingredients of the detergent during the soaking. A bacteriostat in the formulation can eliminate this problem.

The use of bacteriostats will probably increase considerably as soon as a really effective one for use in synthetic granules becomes available. Among the ones most commonly used at present are those listed below. None is as effective as would be desirable, primarily because of insufficient substantivity to fabric under normal washing conditions:

Hexachlorophene

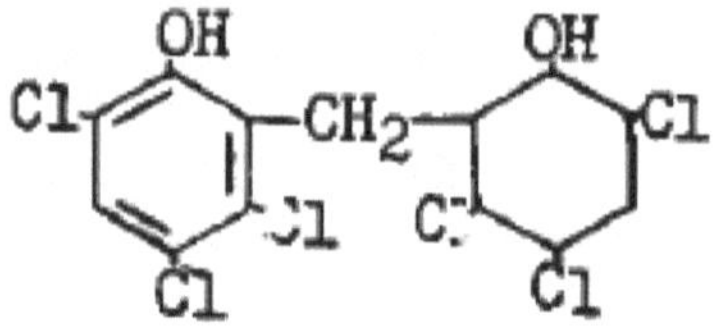

3,+,-trichloro carbanilide (TCC)

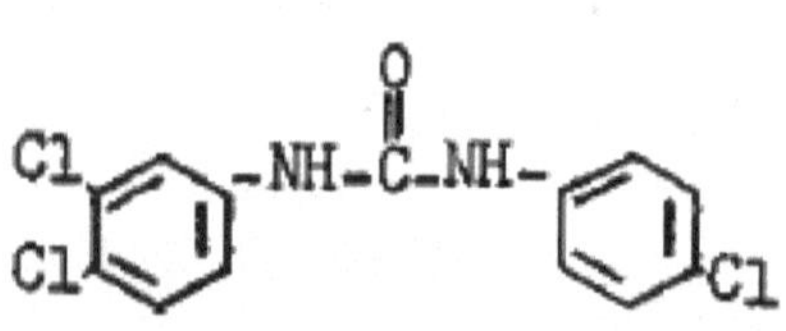

3,3,4,5-tetrachlorsalicylanilide

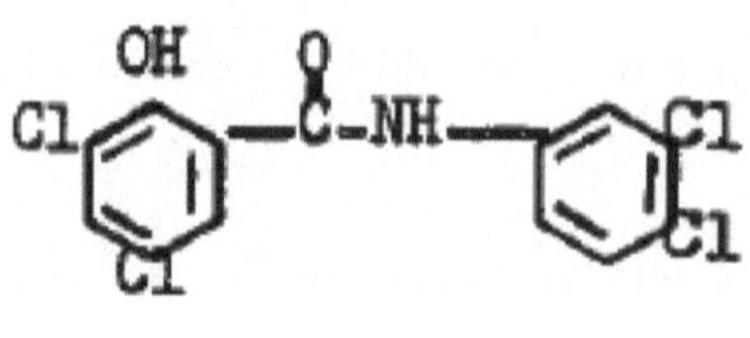

Phenyl mercuric acetate (PMA) (very effective but prohibited in certain countries)

Benzene ring - Hg - OOC - CH:3

Copper sulfate

1.17 SODIUM SULFATE

Sodium sulfate has no beneficial effect on the finished product, serving as a diluent only but a certain minimum amount (about 10% in the base granules) is required for processing reasons.

Sodium sulfate is formed by the neutralization of excess sulfuric acid from the sulfonation reaction. The amount of sulfate formed is fixed by the excess acid required for good sulfonation completeness. When low sulfate levels are required, the excess sulfate or sulfuric acid must be removed by an additional processing step.

Sulfate is a filler in low active, low-cost formulas. Where there is not enough sulfate from neutralization, solid sulfate is added in the neutralization or crutching cycle. Solid sulfate is available as the anhydrous salt and as the decahydrate, and comes in both powdered and granular form. Cost usually decides which form will be used although purity and processing ease are also considered.

1.18 BORAX, SODIUM CARBONATE

Compounds of this type generally do not improve the performance of synthetic granules; they are sometimes used as fillers to cheapen the formula.

1.19 WATER

Water acts as a diluent in the finished product but is important also because it affects the physical characteristics of the product. Too dry a product will be excessively dusty; too wet a product will be sticky and difficult to handle.

Many of the constituents of the finished product can form hydrates. Under any condition of temperature and relative humidity, therefore, each product will have a certain equilibrium moisture content. To avoid humidity caking and changes in outage or density of the packed product on the market, the fresh granules generally are made to the approximate equilibrium moisture content for the conditions under which they are to be sold.

To make finished product of good quality requires raw materials of good quality. Rigid specifications of raw materials have been established for suppliers. Manufacturing Standards outline the precautions necessary to ensure that these materials are not damaged in any way after receipt, during sampling, unloading, and storage.

Most often, raw material receipts which do not meet specifications are rejected. The decision on the final disposition of any raw material is considered to be a plant superintendent or central staff division responsibility. Occasionally, raw materials

TABLE 1.2
Surfactant Production

Surfactant		Production (tons × 1000)	Sales Value ($M)
Amphoteric		50.2	84.6
Anionic		6109.6	1504.2
	Carboxylic acids and salts	1605.2	232.5
	Phosphoric and polyphosphoric acid esters and salts	3269.4	65.7
	Sulfonic acids and salts	1073.4	764.7
	Sulfuric acid esters and salts	103.8	392.4
Cationic		760.2	48.9
	Anine oxides and oxygen containing amines	166.2	738.3
	Amines and amine oxides	96.2	93.9
	Amines not containing oxygen	123.6	80.1
	Quaternary ammonium salts	311.4	327.3
Nonionic		66.8	1417.5
	Carboxylic acid amides	1814.4	138.3
	Carboxylic acid esters	116.6	392.7
	Ethers	1183.8	852.3

which are not greatly off in quality can be used by first blending with material which is of better than minimum acceptable quality. Common raw material defects are identified and listed (Table 1.1), the manner in which these affect the process or the finished product quality and comments on the corrective action, if any, that may be taken with out-of-limit stocks. Note that where the word "economy" is used, the raw material defect is merely one of reduced active level due to inert ingredients. In such cases, the desired level of the material in the finished product can be obtained merely by using more of the raw material and compensating for the inert ingredient by reducing sodium sulfate or moisture level. This should not be done, however, unless the inert material is in no way harmful to the finished product.

1.20 ENZYMES

Enzymes are proteins found in living matter. They catalyze with high specificity biochemical reactions. Enzymes have been used in the preparation of beer and wine. Proteolytic enzymes break down proteins into their components. Some stains (blood, milk, yolk of egg, and chocolate) adhere to the fibers with proteins as binding agent. Proteolytic enzymes eliminate the binding agent and therefore the dirt is easily removed by the surfactant.

Starchy substances can also act as binding agent to adhere proteinaceous and particulate matter to the fabric; amylase enzyme breaks down these starchy links.

Enzymes were dry-mixed in the form of "complex", which meant that the enzyme was embedded in phosphate. In addition, enzymes are added as coated pills (laxatase) or in a marmorized form (alcalase). Several ingredients in the formula influence the performance of the enzymes such as STPP, CMC, Brighteners, etc. Also, pH and the temperature have an effect on the enzyme activity. The pH is optimal at 8.9, the temperature at 50–55 °C. The strength of these proteolytic enzymes is usually expressed in Anson Units (AU).

1.20.1 Analytic Determination

To determine the level of enzymes in the product analytically, the casein method is used. The enzyme is reacted with an excess of a solution of N-Dimethyl casein for a fixed time under controlled conditions of pH and temperature. The amino acids liberated are then reacted with trinitrobenzene sulfonic acid. The resulting yellow color is a measure of the proteolytic activity of the sample and is reported as Z of standard alcalase 0.5 Anson Unit activity.

2 Sulfonation and Sulfation

2.1 DEFINITIONS

Sulfonation is a reaction in which an organic material reacts with a source of SO to form a sulfonic acid. When neutralized, this gives a sulfonate. The reaction takes place at an active hydrogen and the product has a direct C–S linkage (R3–C–SO_3).

Sulfation is also a reaction of a material with a source of SO, but in this case an alkyl sulfuric acid is formed and the neutralized product is a sulfate ester. On sulfation, the reaction takes place at a hydroxyl group so that the product has a C–O–S linkage (R3–C–O–SO_3H).

The chemistry of sulfonation and of sulfation are different and are, therefore, discussed separately below. In addition, in making granular synthetic detergents, the materials which are most commonly sulfonated and sulfated are alkyl benzenes (AB), toluene, and fatty alcohols respectively so that the reactions of only these materials will be considered.

The principal variables in sulf(on)ation are:

The parenthesis in the word sulf(on)ation is used to denote that the statement applies to both sulfonation and sulfation.

(a) Temperature of reaction
(b) Time of reaction
(c) Acid strength
(d) Acid ratio
(e) Effectiveness of mixing

With suitable combinations of these variables, it is possible to fix the following most important reaction characteristics:

(a) Completeness of reaction
(b) Color, odor, and effectiveness of the final product
(c) Ratio of sulfonic (or alkyl sulfuric) acid to HSO in the acid mix acid strength. Thus, the term "acid strength" refers to the SO_4 content of the sulf(on) ating agent.

DOI: 10.1201/9781003596219-3

In this discussion, the term acid will refer only to HSO, oleum, or SO. These are the most often used sulf(on)ating agents for synthetic detergents [chloro-sulfonic acid (CLSOH) has been used primarily in the preparation of actives for liquid detergents].

Acid ratio is a convenient means for expressing the acid usage. It is defined as the weight ratio of acid, as used, to AB or alcohol from a theoretical standpoint; the term is meaningless unless the acid strength and the molecular weight of the AB or alcohol are knowns.

Acid mix refers to the complex mixture obtained from sulf(on)ation. It consists of reaction products, unreacted AB or alcohol, excess acid, side reaction products, etc. Neutralized acid mix is called paste. Completeness is defined as the weight ratio of AB (or fatty alcohol) which has reacted to form AB sulfonic acid (or alkyl sulfuric acid) to the total AB (or alcohol) fed to the reaction mix. It is determined by analyzing neutralized paste for its sulf(on)ated and unsulf(on)ated contents. Completeness cannot be determined by analyzing granules since part of the unsulfurated material is evaporated during spray drying.

Spent acid, normally an excess of acid over the theoretical requirement of SO, is used. In the acid mix, this excess acid is called spent acid and its strength is calculated as percent HSO. It is made up of the SO not consumed in the reaction plus all the water which was introduced with the sulf(on)ating agent. Spent acid is important because, when neutralized, it forms NaSO, which remains in the product. Spent acid strength is a function of acid strength, acid ratio, completeness, and molecular weight of the AB or fatty alcohol.

2.2 SULFONATION REACTION

The desired sulfonation reaction is:

Alkyl benzene + SO_3
Alkyl benzene sulfonic acid R = C to CI5

The reaction takes place at an active hydrogen, primarily in the para position and the product has a direct C-S linkage. The reaction is not reversible.

Sulfonation is a highly exothermic reaction. Efficient heat removal is necessary to avoid charring and to minimize secondary reactions, most of which are favored by higher temperatures. A major factor in the net heat of sulf(on)ation is the heat effect of withdrawing SO_3 from the acid used. Thus, the net heat of reaction increases greatly with increasing strength.

For a given acid strength, the small rise of heat of reaction with increasing spent acid strength reflects increased completeness and the effects of side reactions. With 25% oleum and normal acid ratios, typical heats of reaction are:

97 cals/g AB in sulfonation
80 cals/g tallow fatty alcohol in sulfation
97 cals/g coconut fatty alcohol in sulfation

2.3 COMPLETENESS OF SULFONATION

Since sulfonation is an irreversible reaction, given sufficient time and sulfonating agent the completeness obtained is limited only by the purity of the AB (commercial ABs contain about 1% of unsulfonatable material) and the extent of side reactions. Side reactions except under extreme conditions do not measurably affect completeness. Only rate of reaction and not the maximum degree of completeness obtainable is a function of acid strength.

2.4 COMPLETENESS OF ALKYL BENZENE SULFONATION

In practice, an excess of acid over the stoichiometric requirement is always used for sulfonation so that, neglecting the small effect of side reactions, the actual completeness obtained will depend on the rate of sulfonation. This, in turn, is a function of the following.

2.4.1 Reaction Temperature

Increasing the sulfonation temperature increases the rate of reaction. This effect is small, however, above about 45 °C except when the acid ratios and acid strengths used are such that the spent acid strength drops to about 98% HSO or less.

2.4.2 AB and SO_3 Concentrations

During the initial stages of sulfonation (for example, until about 80–85% of the available AB in a batch system has reacted) rate of sulfonation is a function of the unreacted AB and SO concentrations. Except when sulfonating with pure SO, a considerable molar excess of SO is used. The unreacted SO concentration, therefore, does not change greatly during sulfonation and the reaction rate depends, primarily, on the concentration of unreacted AB.

There are two instances where SO concentration is important. First, there is a critical spent acid concentration (*r* value) below which the reaction does not proceed at any appreciable rate. This is a limiting concentration and the rate is exceedingly slow at this concentration. The value appears to have no theoretical significance in itself especially since it varies with acid strength. Normal acid ratios will maintain the spent acid strength a few percent above the *r* value. The actual acid ratios required to obtain 98% completeness, with normal reaction times, for various strength sulfonating agents with corresponding molar ratios of SO:AB and the resulting spent acid strengths are also shown in laboratory data. Oleum strengths above 30% are not used since poor color and odor result. In addition, concentrations assume a low moisture level in AB or in dilution gas.

Secondly, the concentration activity of sulfuric acid increases sharply at about 100% HSO. The mechanism of sulfonation at low acid strengths (less than about 90% HSO) is believed to be different from the mechanism with oleum (about 100%, and higher).

Experience, however, shows that reaction rate increases sharply in going from an acid strength of less than about 100% HSO, to one above this strength. Beyond about 85% completeness, the reactant concentrations begin to decrease rapidly and so does the sulfonation rate. At this point, degree of mixing becomes increasingly important, presumably because very good mixing of the viscous acid mix is necessary to bring the two reactants, AB and SO_3, together for reaction. Thus, in normal operation, in the completeness range of interest (97–98%), mixing, rather than reactant concentrations, controls the overall reaction rate.

2.4.3 Degree of Mixing

The importance of effective mixing is illustrated by the data:

Mixing (rpm)	Completeness (%)
900	92.5
1250	93.8
1750	95.5
2500	97.9

These data are for a special continuous sulfonator with no external reaction coil and are relative only. Increasing speed of mixing may be taken as better mixing; the effect on completeness is evident.

Side reactions (color and odor of the final product) and several secondary reactions also occur during sulfonation:

(a) Di-sulfonation probably occurs slowly by reaction between SO and the alkyl benzene sulfonic acid formed in the primary reaction. Although this reaction is believed to take place, it has not been confirmed by paste analyses. The amount of di-sulfonate formed is certainly small and is probably unimportant.
(b) Oxidation and dehydration reactions take place largely under conditions of high SO concentration and high temperature. An example of the latter is sulfone formation. The removal of water by SO is the driving force in this reaction. Most other side reactions are also favored by high SO concentrations; consequently, it is preferable to add acid to AB rather than vice versa.
(c) Rearrangement of the alkyl chain probably occurs to some extent during sulfonation, but this has not been studied in detail.

Many of the factors which increase the rate of sulfonation also increase the rate of the side reactions. As previously noted, side reactions reduce the completeness obtainable. In addition, however, side reaction products may cause unsatisfactory color and odor even when present in such small amounts that their effect on completeness is negligible. As with sulfonation, the rate of the side reactions decreases rapidly as the acid strength drops below 100% HSO presumably again, because the activity of the sulfuric acid drops at this point. When 20–30% oleum is used as a sulfonating agent, the maximum allowable sulfonation temperature for good color is approximately 5

°C, whereas with 100% H_2SO_4 the acid mix is far less sensitive to color degradation and temperatures up to about 60–70 °C are used.

Although the mechanism of color formation is not fully understood, it probably results from side reactions which occur at the surfaces of the acid droplets formed upon injection of the acid into AB or acid mix. Since the reactions are exothermic, local overheating at the acid droplet surfaces accelerate the charring and other undesirable side reactions. Good mixing to dissipate the high local concentrations of SO and to prevent high local temperatures is of major importance in minimizing color formation. Effectiveness of mixing determines, to a large extent, the maximum acid strength which can be used without obtaining poor colors. The better the mixing, the higher is the acceptable acid strength.

Under normal conditions of acid ratio, acid strength, temperature, and mixing side reactions are generally not a problem. Difficulty with color can occur, however, if any of these factors is out of control. This happens most often with acid ratio or temperature during startups or shutdowns.

The desired sulfation reaction is:

$$ROH + \quad SO_2R\text{–}O\text{–}S\text{–}OH$$

Fatty alcohol Alkyl sulfuric acid

The reaction takes place at a hydroxyl group so that the product has a C-OS linkage. In the presence of water, this reaction is reversible by hydrolysis. Sulfation with sulfuric acid or with oleum is an equilibrium reaction, the maximum possible completeness being limited even without consideration of side reactions. However, because of the reactivity of fatty alcohol, the side reactions are more numerous, more serious, and more rapid than in sulfonation. These are, again, dehydration and oxidation reactions. Some typical side reactions are:

$$RCH,CHOH + SO_3$$

$$2RCH,CHOH + SO_3$$

$$2RCH,CHOH + SO_4$$

$$RCH,CHOH + SO_4$$

$$RCH = CH + HSO$$

$$RCH,CH,OCH,CH,R + H,SO$$

$$COCCH_3R + 2SO_3 + 2SO_2$$

The un-sulfated portion of an alkyl sulfate paste, obtained by extraction with petroleum ether, was analyzed with the following results:

Component	Unsulfated Portion
Un-saturates	3.8
Di-alkyl ether	4.7
Ester	5.1
Fatty alcohol	1.3
Aldehyde	0.2

The incompleteness is primarily what may be called over-sulfation rather than under-sulfation. The term "un-sulfated" is, however, commonly applied to this portion of a paste.

Although the sulfation reaction is very rapid, the alkyl sulfuric acid is not stable. Both the alcohol and the alkyl sulfuric acid undergo degradation reactions with oleum. The net effect is that the completeness, as measured by the amount of alkyl sulfuric acid obtained, has a maximum at a rather short reaction time.

Optimum conditions for sulfation are based upon obtaining high primary reaction completeness and terminating the reaction to minimize the secondary reactions.

The sulfation reaction is terminated by neutralizing the acid mix since alkyl sulfate is relatively stable at a pH above 7. Since alkyl sulfuric acid hydrolyzes in the presence of water, neutralization is usually accomplished by adding acid mix to the neutralizing agent (NaOH or $NaCO_3$) instead of vice versa. The heat of neutralization must be removed to avoid decomposition of the alkyl sulfuric acid.

Sulfonation is carried out in both batch and continuous systems. Sulfation is generally a continuous process, although batch systems are used when SO or chlorosulfonic acid is the sulfating agent.

2.5 BATCH SULFONATION

A typical batch sulfonation unit consists of a vessel equipped with a means of mixing the AB and acid and of removing the heat of reaction. Tanks with cooling coils or jacket, for example, and an agitator may be used. A more effective system, however, consists of a vessel with an external heat exchanger through which the acid mix is circulated by means of a recirculation pump.

All the AB is charged to the tank and recirculated through the cooler as acid is slowly added. The acid is conveniently added at the pump, where most of the mixing takes place. An agitator is generally also provided in the vessel but serves principally to maintain homogeneity within the vessel. Heat of addition of acid is slow (30–60 minutes per batch) to ensure good mixing of the acid as it is added and to minimize the load on the cooling equipment. After all the acid has been added, an additional aging period (30–60 minutes) is required to allow the sulfonation to go essentially to completion.

In such batch systems, mixing tends to be relatively poor. Completeness is often lower than is obtainable by other processes and the finished product colors are somewhat poorer. The batch system has the advantages, however, of requiring the cheapest and simplest equipment and providing the greatest ease of operation with a minimum of process control problems.

There are no metering difficulties and few equipment problems. Actual rate of acid addition is not important provided only that the maximum permissible rate is not exceeded. Total amounts of AB and acid added are readily controlled and so is the temperature.

2.6 SEMI-CONTINUOUS SULFONATION

The relatively poor paste color and completeness obtained in batch sulfonation generally arise from two factors:

(A) POOR MIXING

In a batch system, the acid passes through the circulation pump initially, then must reside in the tank for more than a minute before it reaches the mixing action of the pump again. Most of the mixing of ingredients in this system occurs in the pump, very little is due to the tank agitator.

(B) REACTANT CONCENTRATIONS

In the early stages of a batch sulfonation cycle, acid is added to pure AB. Much of the sulfonation, therefore, occurs by the direct action of high strength acid on the AB. This condition promotes side reactions and poor color. Not until later in the course of sulfonation are ABS and spent acid present in large enough quantities to disperse and dilute the AB and acid.

A so-called "semi-continuous" sulfonation provides improved reaction conditions over those usually obtained in batch systems. The only additional equipment needed is flow meters for AB and acid and a by-pass line from the discharge of the cooler to the suction of the recirculation pump. Both AB and acid are metered manually into the sulfonator pump. The acid mix formed is recirculated through the dominant bath consisting of the pump, a cooler, and the tank by-pass line.

Initially, the sulfonation tank is empty but, during the course of the sulfonation run, it gradually fills with the acid mix from the dominant bath. After each batch the lines are normally left full of acid mix. To drive out any possible air pockets which would cause pump cavitation and lack of proper mixing, some AB, or acid mix from the preceding batch, is run into the system at the start of each batch until the external system is completely full of liquid. To minimize the side reactions which result from high SO concentrations, the rate of acid addition is set to give a slightly lower acid ratio than is finally desired. This makes control of the addition rates less critical by reducing the danger of instantaneous over-concentration of SO.

As a result of this precaution, the AB addition is completed first and the acid addition is then completed as in normal batch sulfonation. That is, the acid mix is recirculated through the cooler and directly back to the tank during addition of the final portion of acid. A 15–30 minutes aging period for the acid mix then follows.

In the semi-continuous procedure, most of the reaction occurs in the dominant bath. Mixing is greatly improved and, since both AB and acid are quickly dispersed in the recirculating stream of acid mix, sulfonation occurs, to a large extent, by the action of spent acid on AB. These improvements increase completeness by as much as 2% over the batch system and noticeably improve color.

2.7 CONTINUOUS SULFONATION (DOMINANT BATH)

Heat removal and mixing are the major problems involved in making the sulfonation process continuous. Both difficulties are readily solved by a dominant bath process. In a dominant bath process, the reactants are injected into a large mass of reaction products, the relative magnitude of the latter being sufficient to be predominant. The heat of reaction is thus dissipated into the large mass of material without excessive temperature increase, and the larger volume facilitates heat removal and mixing.

Various types of dominant bath systems are used. A common dominant bath system is the continuous stirred-tank reactor (CSTR). Any batch unit can be converted to continuous use by providing accurate metering equipment for the AB and acid and a suitable overflow for the acid mix. The Ballestra sulfonation systems are of this type. The rest of this discussion, however, concerns the custom design of dominant bath, which is both simpler and more efficient.

In the custom design, the reaction products (acid mix) are continuously pumped in a dominant bath loop consisting essentially of a heat exchanger, piping, and a pump. AB and acid are metered to the pump as in the semi-continuous system and acid mix is continuously removed. Here again, most of the mixing occurs within the pump. A back-pressure control valve maintains a positive pressure on the system at all times to prevent air entrainment and pump cavitation.

In "ideal" dominant bath, all parts of the bath are at the same composition, the composition of the effluent stream. The mean reaction time is equal to the system volume divided by the rate of volume throughput. In an actual dominant bath loop, the "recirculation ratio" (the flow rate of recirculated material divided by the flow of reactants) determines the degrees of approach to ideality. For example, at a 5:1 ratio, one-sixth of the feed will be removed the first time past the take-off point, thus short circuiting the system and having a reaction time considerably less than the calculated mean. However, at a recirculation ratio greater than about 5:1, departure from ideality is negligible.

If the chemical reaction rate is controlling, the rate is proportional to the concentration of unreacted materials. In a batch reaction, the reaction is fast initially since reactants are high in concentration. Then, as the reactants are consumed, the rate decreases. In a dominant bath, the reactants are always at the final concentration, and hence the chemical reaction is relatively slower.

This fundamental feature of the dominant bath system may be seen more clearly by considering first the batch reaction curve; the instantaneous reaction velocity is as discussed above since it is dependent upon the concentration of the reactants. This instantaneous reaction velocity is the slope of the curve, that is, the change in completeness with time. The average reaction velocity would be expressed by the slope of the line connecting any point on the curve with the origin. However, in an ideal dominant bath, all reactants are at the final concentration so that the reaction velocity would be expressed by the slope of the batch reaction curve at the particular completeness being considered. A line from the origin and parallel to the tangent of the batch reaction curve at that particular completeness would give the average reaction velocity or the effective reaction velocity in the dominant bath. Where this line reaches the level of completeness being considered, the reaction time may then be noted by referring to the abscissa. It is obvious from this description that at any degree of completeness, a considerably greater reaction time is required in the dominant bath and that the higher the completeness being considered, the greater the difference in reaction time required for dominant bath as compared to batch reaction. In this description, the term "straight through reaction" has been used and is indicated to be the same as the batch reaction. This term is applied to a continuous process in which there is no end-for-end mixing, and thus each increment of the mixture may be considered as a very small batch.

While the problem of heat removal is readily handled by means of the dominant bath system, the reaction time requirements for high completeness in a dominant bath may be excessive. Thus, a combination of dominant bath and straight through sections is advantageous.

The required reaction time in the dominant bath (and hence the volume) can be reduced by operating at a lower completeness since the concentrations of reactants will then be higher. This is feasible if the desired final completeness is obtained by providing additional reaction time after the dominant bath. This is the function of a reaction coil. The coil minimizes end-to-end mixing, and thus we have the straight-through reaction which has, as stated above, the same reaction kinetics as a batch. The temperature rise in the coil is generally small since most of the reaction and cooling has already occurred in the dominant bath.

The importance of the division of reaction time between dominant bath and coil (straight through section) is demonstrated. The abscissa is the total reaction time, dominant bath plus straight through. The points on a curve represent the dominant bath completeness at a given reaction time. Sequential curves continuing from these points would show the progress of the reaction in the subsequent straight through sections. These curves are typically for an experimental sulfonation unit. Actual plant systems give higher completeness due to the better mixing obtained on the larger units.

Toluene sulfonic acid is made by sulfonating toluene with oleum in the same system with the alkyl benzene. Toluene is a short chain alkyl benzene, having one methyl group attached to the benzene ring. Since its molecular weight is lower than regular alkyl benzene, it will take up considerably more SO than will an equal weight of AB. Furthermore, toluene is more reactive than AB and, in a mixture of the two (co-sulfonation), is sulfonated preferentially, so that any deficiency in acid results in a poorer completeness for the AB than for the toluene. Toluene, however, is readily sulfonated by 98–100% HSO. Consequently, when minimum oleum usage is desired (for either economic or formulation reasons), a "series" sulfonation of AB and toluene can be utilized. The AB is sulfonated first using all the acid. The toluene is then added. The toluene completeness may not be as high as with co-sulfonation, but since un-sulfonated toluene is almost entirely evaporated in the spray tower, it will not cause the stickiness and performance problems that occur with high un-sulfonated AB. Series sulfonation of toluene can be carried out in either the batch, semi-continuous, or continuous units.

In the batch or semi-continuous units, toluene is added after the AB aging period. In the continuous system, a separate sulfonator is required.

The following data illustrate the advantages of series sulfonation using 25% oleum:

	Co-sulfonation	Series Sulfonation
Acid ratio	1.2	1.05
AB completeness	97%	98%
Toluene completeness	98%	85%
Spent acid strength	98%	95%

Note that toluene completeness drops, while AB completeness remains high. This shows that toluene will not be sulfonated by alkyl benzene sulfonic acid. Minimum acceptable acid ratios for sulfonation of toluene is limited by the requirement that toluene completeness not be too low. Below 80%, the residual un-sulfonated toluene contributes a noticeable odor to the finished product.

The reaction time required for toluene sulfonation is less than that for AB. Toluene co-sulfonation, when used, is carried out under the same reaction time and temperature conditions as for AB alone (3 minutes minimum in the external coil, 5 °C, maximum).

With series sulfonation, the acid available for toluene sulfonation is only 98–100% HSO instead of being oleum as in co-sulfonation. Consequently, the rate of toluene sulfonation is lower. The minimum reaction time in the toluene external coil must, therefore, be increased (minutes minimum). Since the rates of the various sulfonation side reactions are similarly reduced, series sulfonation can be performed at higher temperatures (up to about 65 °C). The higher temperature increases the rate of toluene sulfonation but does not hurt product color or odor. The higher allowable temperature also permits substitution of an in-line mixer for the entire toluene dominant bath system, since toluene sulfonation raises the acid mix temperature only about 10 °C (from about 50 °C to about 60 °C).

Toluene is more susceptible to side reactions than is AB. Ortho and meta toluene sulfonic acids, di-sulfonic acids, and di-toluo-sulfones may be formed. These compounds can cause stickiness in the spray-dried granules. Their formation is prevented by keeping the spent acid strength below 99% HSO.

2.8 CONTINUOUS SULFATION: DOMINANT BATH

The continuous dominant bath loop, although not ideal, is the only practical means of providing the efficient cooling and short reaction time required for sulfation of fatty alcohols with oleum. (Sulfations with SO_3 or chloro-sulfonic acid, however, are generally performed in a batch reactor.) Equipment limitations still impose restrictions upon minimum reaction times. Further, the high concentration of alkyl-sulfuric acid and low concentrations of alcohol in the dominant bath favor the secondary reactions relative to the desired reaction.

Equilibrium completeness calculated shows the importance of side and secondary reactions from data obtained under mild reaction conditions such that side reactions were negligible. Equilibrium is not attainable in practice except at very low spent acid strengths. Data in the low strength spent acid region indicate time is insufficient for attainment of equilibrium. With high strength spent acid, the increasing speed of secondary reactions again widens the spread. Above a critical spent acid concentration, the secondary reactions reach completeness.

Coconut alcohol sulfation consumes the alkyl-sulfuric acid faster than it is formed, and so completeness reaches a maximum. This is further illustrated by the reaction completeness as determined by disappearance of fatty alcohol and as measured by the formation of alkyl sulfuric acid. The difference between the two curves is a measure of the loss of completeness due to secondary reactions. Within practical limits, lowering the reaction time increases net completeness at any given temperature by minimizing secondary reactions. This is illustrated by experimental data. Reaction time becomes more important when the spent acid strength is higher because of the acceleration of the degradation reactions.

Lowering the temperature also improves completeness for the same reasons; therefore, operation at the lowest obtainable temperature is desirable. The loss of about 2.9% completeness for every 10 °C increase in sulfation temperature holds in the range of 3–5 minutes reaction time but probably decreases to about 1.1% per 10° C at one minute reaction time. Within practical limits, lowering the reaction time increases net completeness at any given temperature by minimizing secondary reactions.

The minimum operable temperature is, of course, limited to a temperature above that at which the alcohol starts to solidify. The relationship between time and temperature is best explained with reference to hypothetical contours of sulfation completeness as a function of time and temperature. Operation in the prohibited region is due to solidification of the alcohol. Operation is also prohibited due to the limitation of equipment, that is, the dominant bath volume is too large and the heat exchanger area is too small to sustain the temperature at higher rates (shorter reaction times).

Contours of sulfation completeness in a dominant bath for varying time and temperature are available at constant acid ratio. A separate set of contours with its own boundary limitations exists for each fatty alcohol. Further, for each fatty alcohol, the contours may shift slightly for changing acid ratio thus generating a solid response surface. Each system, naturally, has its own optimum point of operation within the given boundary limitations. The high freezing points of fatty alcohol and acid mix also present problems in heat transfer. These materials tend to freeze on heat exchange surfaces if the cooling water is too cold. Water above 25 °C is safe, although, under certain conditions still warmer water may yield lower reaction temperatures due to the proportionally larger increase of the heat transfer coefficient as compared with the decrease of the temperature difference between product and water.

2.9 CONTINUOUS SERIES SULFATION

Series sulfation was developed as a means of improving fatty alcohol sulfation in those formulations using a mixed active of ABS and AS. It reduces the reaction time for sulfation while maintaining a reasonable total volume in the dominant bath for effective heat transfer. AB is sulfonated alone first and the acid mix obtained is then pumped through a second dominant bath system, where fatty alcohol and additional oleum are added. This reduces the volume available for the AS acid mix. The ABS acid mix is not affected by this additional operation. Besides reducing AS reaction time, the ABS acid mix acts as a thinning agent and thus permits operation at lower temperatures with fatty alcohol. The dilution effect allows use of higher acid ratios in the AS system. All of these effects improve sulfation completeness.

Series sulfation introduces an additional process variable, the division of the total acid usage between the ABS and AS systems. Since the AB sulfonation is independent of the subsequent sulfation, the optimum acid ratio for the AB in the series operation is the same as it would be in simple sulfonation. The acid ratio for the AS sulfation, however, can be higher in the series operation because the degradation reactions are repressed. If low total acid usage is necessary to reduce sodium sulfate in the final product, the AS acid ratio should be decreased at constant ABS acid ratio because ABS completeness falls off more rapidly with insufficient acid than does AS completeness. This insures highest overall completeness.

Both ABS and alkyl sulfates can be made by direct reaction of AB or fatty alcohol with SO. Since no water is present to dilute the acid as the SO is used up by the sulf(on)ation reaction, a large excess of acid is not required as is the case with oleum. Normally, 100–105% of the theoretical SO is used. The pastes that result upon neutralization will, therefore, have a very low sodium sulfate content.

Generally, liquid SO is too reactive to be used directly. One way to render it less reactive is by diluting it, in the liquid phase, with some inert solvent (e.g., SO under pressure), but this method is generally too expensive, since the solvent must be recovered. A simpler method is to use SO vapor diluted with 10–13 volumes of dry air or other inert gas. The diluted gas mixture is obtained by one of two ways:

(A) Sulfur Burning

The process is the same as that used in the manufacture of oleum. A metered quantity of sulfur is burned with excess air in a furnace and the resulting mixture of air and SO is passed over a catalyst at high temperature to effect the conversion of SO. The resulting gas mixture usually contains 8–10% SO_3 and is used directly for sulf(on)ation. The air must be dried thoroughly or the moisture will form an acid mist with the SO_3 which reduces the amount available for sulf(on)ation. A major disadvantage of this process is that, because of the high temperatures required for sulfur burning and SO conversion, it is uneconomical to start up and shut down the process frequently. A detergent plant using this system must operate around the clock, at least in the sulfonation stage, unless the mixture can be diverted to an oleum making plant.

(B) Vaporization of Liquid SO3

Pure sulfur trioxide is known to exist in three physical forms, of which the one that is liquid from 17 °C to 45 °C, at atmospheric pressure, is the least stable. The two polymeric forms are more stable but are undesirable because of their higher melting points. By the addition of small amounts of certain compounds, known as stabilizers, the SO_3 can be kept in the liquid form. Even when stabilized, great care must be taken to exclude moisture, because the presence of even small amounts of water can cause the liquid form to revert to one of the solid forms. For use in sulf(on)ation, the liquid SO is metered into a vaporizer, and the resulting vapor is mixed with the proper amount of dry air.

A new process, which has been applied to liquid detergent manufacture, is to run the sulf(on)ation under vacuum using vaporized but undiluted SO3. Results are similar to those obtained using an air-SO_3 mixture having the same partial pressure of SO3.

The main requirements for good, finished product quality are close control of the acid ratio, intimate mixing of the gas with the AB or alcohol, and close control of the reaction temperature. The sulf(on)ation process can be either batch or continuous; however, batch processes have thus far yielded products of lighter color. The batch process is similar to that described for sulfonation with oleum except that the mixing is done in the tank rather than in an external pump. The choice between continuous and batch process will depend on the economics of a given project and also on the finished product color requirements.

The intimate mixing required is obtained by the use of high-speed turbine agitators and/or injection of the gas at high velocity through small jets. The gas should preferably be introduced near the tip of the turbine. Reaction temperature is maintained below 60 °C, by cooling jacket and coils or by use of an external pump-heat exchanger arrangement. The design of cooling and mixing equipment for AB sulfonation is difficult since the final acid mix is very viscous (700 cp at 60 °C).

A peculiar effect is noted upon neutralization of finished ABS acid mix in an alkaline seat. Even though the pH is adjusted to neutral or slightly above, it shifts towards acid in a few hours. This pH drift may continue for some days. This is explained by the formation of an acid anhydride in the acid mix which hydrolyzes slowly in an alkaline seat. The problem is avoided by adding 1–2% water to the acid mix with good agitation and cooling at the end of the SO3 addition and prior to neutralization. The anhydrides hydrolyze rapidly under these conditions and problems of pH drift are eliminated.

2.10 CHOOSING ACID STRENGTH AND ACID RATIO

The optimum acid strength and acid ratio will depend on the finished product formulation, the relative costs of acid, caustic, sodium sulfate, and the process equipment available. No hard and fast rules can be given since costs of raw materials vary widely from one country to another; however, in most cases, the strongest available oleum of satisfactory quality (up to 30%) should be used if the sulfonation equipment can handle it. Since 30% oleum freezes at about 21 °C, compared to 27 °C for 25% oleum, adequate tracing and insulation must be provided if 30% oleum is contemplated.

The choice of oleum strength fixes the acid ratio within rather narrow limits. For example, using 25% oleum, acid ratios for AB sulfonation can vary between about 1.00 and 1.15. Completeness will vary from about 97.5 to 98.5, and the weight ratio of ABS to Na_2SO_4 in neutralized paste will vary from 1.1 down to 1.16. Lower acid ratios hurt completeness, and higher acid ratios hurt color and odor. The active to sulfate ratio is, of course, affected when toluene is also sulfonated. With co-sulfonation of toluene, higher acid ratios can and must be used (up to about 1.20), and the active to sulfate ratio will tend towards the low side. With series sulfonation of toluene, the minimum acid ratios can be used, and active to sulfate ratio can be increased above 1.1 since toluene tends to use up the spent acid. Higher active to sulfate ratios can only be obtained by using stronger oleum or by using the acid mix concentration process.

2.11 PROCESS CONTROL

To make a finished product, meeting all the specified limits requires careful control at each process step. In particular, sulfonation conditions are not only important to the ease of operation of subsequent steps but can also directly affect the color, odor, and composition of the finished product obtained. Following is a discussion of some of the more common process control problems.

2.11.1 Mixing Efficiency

Effective mixing of AB or alcohol prevents high local SO concentrations and high temperatures. High SO concentrations, persisting for even a fraction of a second, hurt color and odor and in extreme cases hurt completeness. Injection of oleum through a nozzle into the eye (suction) of a centrifugal pump was mentioned as one way of obtaining good mixing. Still better mixing can be obtained by the use of peripheral injection or "Venturi" mixing.

Peripheral injection of oleum involves modification of the sulfonator pump to inject acid into the pump through nozzles at the impeller tips. Since this is a point of highest shear, mixing of the acid with re-cycled acid mix is greatly improved over eye-of-the-pump injection. Higher acid feed pressures are, of course, required.

Important mechanical features of peripheral acid injection include acid injection nozzles and O-rings, which must be properly installed and maintained to prevent leakage of acid past the O-rings. Such leakage would defeat the purpose of the installation.

Venturi involves the installation of a special mixing nozzle in the dominant bath loop. The details of the Venturi and how it is installed in the dominant bath are important to this application. The word Venturi is a misnomer since the device has no gradually expanding section such as in a Venturi meter. The mixing action is due, in fact, to the sudden expansion of acid mix which produces violent cavitation. Oleum and AB are injected at the throat of the Venturi. The major disadvantage is that the recirculation pump must provide additional pressure to force the acid mix through the Venturi as well as through the heat exchanger and piping.

2.11.2 Venturi Mixer Sulfonation

These improvements in mixers give improved colors and completeness on pastes made from 20% to 25% oleum and make possible the use of 30% oleum for sulf(on) ation. Color and completeness are at least as good as normally obtained with 25% oleum and eye-of-the-pump injection and, in some cases, have even been better.

2.11.3 Metering

As previously noted, sulf(on)ation completeness, color, odor, and active to sulfate ratio are highly dependent on acid ratio. The proper proportion of reactants must be metered into the system within fairly close tolerances if all finished product limits are to be met. Calibrations of metering equipment and checks on acid ratio by a specific gravity test are used. This test is based upon the fact that over the range of ratios in use, a mixture of oleum and alkyl benzene has a specific gravity that is a linear function of acid ratio.

In the batch or semi-continuous process, the instantaneous acid ratio must never exceed the final acid ratio. Further, the total quantity of AB and acid must be correct since each batch of AB will be processed further into finished product and any errors in the total quantity of AB and/or acid will show up as deviations from the desired finished product composition.

In the continuous process, the instantaneous acid ratio must always be controlled within fairly tight limits. However, the feed rate of all materials can be changed whenever there is an excess or deficiency of paste available for crutching. This, of course, assumes a batch crutching operation.

In either process, an excess of un-sulfonated alkyl benzene in the finished product can reduce sudsing performance and can cause handling and packing problems due to stickiness.

Toluene metering must also be carefully controlled, particularly if toluene is co-sulfonated with AB. Too much toluene, besides diluting the formula, lowers AB completeness since it sulfonates preferentially to AB. Too little toluene would have the same effect as too high an acid ratio such that poor color and odor might result. In series sulfonation of toluene, AB completeness is unaffected by errors in toluene metering; formula deviations would, of course, still occur. Too much or too little sodium toluene sulfonate in the finished product will affect the ease of processing and change the caking resistance of many formulas. Low STS content can usually be detected by increases in paste and crutcher mix viscosity.

2.11.4 Mix Viscosity

High acid mix viscosity can be the result of no toluene, too little spent sulfuric acid, too low an acid mix temperature, or air entrainment. High viscosity will generally result in poor mixing, bad color, and low completeness. A compromise must be reached between a sulfonation temperature which is high enough to give a low viscosity acid mix and yet low enough to avoid development of poor colors from temperature alone.

2.11.5 Viscosity of ABS / Acid Mix

Some of the adverse effects of high acid mix viscosity can be guarded against by ensuring that the sulfonation equipment is always in ideal working order. Pumps must be inspected periodically for clearances, and the location and condition of stock nozzles must be frequently checked.

Air entrainment through the pump packing gland and air from stock lines can result in an emulsified acid mix which is highly viscous and difficult to pump. In the extreme, complete cavitation of the pump can occur, resulting in no mixing at all. If this happens, it is normally necessary to shut down the system and purge with AB, venting the air from the top of the heat exchanger.

An important side effect of the viscosity of the sulfonic acids is the decrease in performance of heat exchangers at low temperatures. With very cold water, the heat transfer may actually be reduced due to the formation of a viscous film on the heat exchanger surface even though the bulk of the acid mix is above 50 °C. Cooling water temperature above 25 °C is safe.

2.11.6 Recirculation

Recirculation rate is important because of its effect on both mixing and temperature. Low recirculation tends to give poor mixing and high temperatures, both of which hurt color and completeness. The best indication of insufficient recirculation is an increased temperature drop across the heat exchanger. Since total cooling load is a function of the reaction occurring within the system, it may be assumed constant for any given feed rate. Any increase in the difference between acid mix inlet and outlet temperatures at the heat exchanger can be assumed to be the result of decreased circulation. Poor recirculation can result from any of the factors causing an increase in acid mix viscosity. In addition, poor recirculation can also result from a plugged or fouled heat exchanger. This is not too common with AB acid mix but results frequently when very cold cooling water is used with fatty alcohol acid mixes.

2.11.7 Reaction Temperature

Excessively high reaction temperature can result from a variety of causes. The most common one is inadequate recirculation, as noted above. In addition, insufficient cooling water of too high a temperature is a common cause. Gradual increases in rate above design rate for the system must, of necessity, push reaction temperature up during the hotter months of the year. The product negatives resulting from high temperatures are bad color and odor in the finished product. Sulfonation completeness will normally not be affected unless temperatures are above 55 °C and charring results. As previously noted, maximum sulfation completeness is attained at the lower temperatures.

2.11.8 Reaction Time

Insufficient reaction time in either the dominant bath or the external coil generally results in decreased sulfonation completeness. Since reaction time is normally a function of the design of the system, there is little that the factory can do to change conditions. Typically, manufacturing standards specify a minimum reaction time of 0.6 minutes for the dominant bath and 3.0 minutes for the external coil. The first figure will not normally be approached, simply due to the geometry of the systems.

The 3-minute minimum for external coils must not be violated as production rates are increased.

2.11.9 Back Pressure

The back pressure on the dominant bath is controlled by automatically opening or closing a valve in the outlet from the system. The back pressure is sensed at the suction of the recirculating pump, and a signal is transmitted to the valve by a controller. Response is normally good as long as the control valve is properly sized and the sensing element functions properly. Three methods for sensing pressures are normally used:

(a) Sensing the pressure of the organic stock.
(b) Bleeding a small portion of the organic stock through a secondary inlet line.
(c) Sensing directly from the acid mix stream to a diaphragm.

While (c) is slightly more costly to install, it is usually justified in the long run due to freedom from sensing difficulties.

The acid mix take-off can be at any point in the dominant bath but is usually either at the suction or the discharge of the recirculation pump. The advantage of suction-side take-off is that only one control valve is required no matter how many dominant bath loops are in series. In this case, the control valve is placed in the take-off line of the final dominant bath. With discharge-side take-off, a separate control valve is required for each loop.

2.11.10 Water Contamination

A problem which is not too common, but which is very serious when it does occur, is that of water entering the acid mix system, either with the raw materials or by leakage through the heat exchanger or pump packing glands. This water can dilute the oleum and cause extremely low sulf(on)ation completeness.

3 Acid Mix Concentration

In order to obtain an adequate rate of sulfonation, the acid strengths and acid ratios normally used are such as to provide a considerable excess of SO over the stoichiometric requirement. This excess of sulfonating agent, along with any water present, constitutes the spent acid. In the neutralization step, spent acid is converted to sodium sulfate and water. The sodium sulfate thus formed dilutes the finished product and may limit formulation flexibility. In particular, it limits the weight ratio of ABS to sodium sulfate to the range of roughly 1/1 to 2/1 when sulfonating with sulfuric acid or 30% oleum, respectively. However, with the need for a high active formula (e.g., 35% ABS to 35% STPP) for use in overseas countries, active to sulfate ratios of 3/1 or higher are required. The acid mix concentration process provides a relatively simple means of increasing the active to sulfate ratio.

Typical acid mix resulting from the sulfonation of alkyl benzene with oleum is an intimate mixture of spent sulfuric acid (about 98% HSO) and alkyl benzene sulfonic acid. Concentration of the acid mix is accomplished by separating and removing a part of the spent acid before neutralization. This may be done either by centrifuging or by settling. However, centrifuging is not practiced to any great extent commercially because of corrosion problems, and common practice is to settle the acid mix either in a batch system or continuously. The lower layer obtained from settled acid mix is spent sulfuric acid which can be drawn off and recovered. The upper layer is concentrated sulfonic acid which can be neutralized to form a paste with an active to sulfate ratio as high as 9/1.

3.1 FACTORS AFFECTING THE DEGREE OF SEPARATION

Dilution

The degree of separation obtained at equilibrium and, thus, the active to sulfate ratio obtained from any concentrated acid mix is a function of the spent acid strength. Although some separation may be effected at 97–98% spent acid strength, dilution of the acid mix with water is necessary to obtain more complete separations. The degree of separation increases with dilution reaching a practical limit at 77–78% HSO. Spent acid strengths in the range of 80–90% and below 68% cause maximum corrosion and

DOI: 10.1201/9781003596219-4

therefore operation in these regions should be avoided. Normally, there is no reason to operate below 77% HSO.

3.2 SPENT ACID STRENGTH, PERCENT HSO

Dilution of acid mix is highly exothermic. The heat evolved corresponds essentially to the heat of dilution of the spent sulfuric acid present in the mix. The temperature of the diluted acid mix determines the settling temperature and is, therefore, critical:

(a) low temperatures will result in low settling rates.
(b) high temperatures will result in color and odor degradation of the acid mix and also in an increase in its corrosiveness. Adequate cooling capacity must, therefore, be provided for the dilution step.

3.3 SPENT ACID

The acid mix, after dilution to 78% spent acid strength and before settling, is a mixture containing parts of spent acid to parts of sulfonic acid. In this mix, the sulfonic acid is the continuous phase. It is a relatively viscous phase through which the heavier droplets of spent acid move downward fairly slowly. However, if sufficient spent acid is added or recycled to the acid mix feed to change the ratio of spent acid to sulfonic acid from 3:4 to about 6:4, the system will change from a continuous-sulfonic to a continuous-sulfuric acid phase. Since the spent acid is merely recycled, it does not affect the overall material balance across the acid mix concentration process. However, the continuous sulfuric acid phase produces a much less viscous acid mix through which the lighter droplets of sulfonic acid move upward rapidly. This effect is predicted by Stokes Law which says that the terminal velocity of a falling (rising) particle is inversely proportional to the viscosity of the continuous phase. The end result of spent acid recycle is increased rate of settling with no detrimental effect on product quality.

3.4 SETTLING TEMPERATURE

Another way to reduce the viscosity of the continuous phase, and thus increase the rate of settling, is to settle at a higher temperature. When spent acid recycle is used, the viscosity of the continuous (spent acid) phase is already so low that increasing the temperature above about 50 °C has little effect.

The final equilibrium point reached does not change significantly with temperature. An increasing rate of corrosion and a deterioration of product color limits the maximum settling temperature to about 55 °C.

3.5 SETTLING TIME, MINUTES

In a batch system, maximum rate of settling will be obtained with a vessel in which the diluted acid mix is in the thinnest possible layer so that the components of the two layers have the shortest possible distance to travel.

In continuous acid layering systems, the relationship between increased rate of settling and decreased height of the feed mix is complicated by another factor. For thin layers, flow turbulence effects become important so that, if the layer becomes too thin, an excessive amount of spent acid may be entrained. In addition, the flow rate affects the degree of approach to equilibrium. Thus, the greater the upward velocity of sulfonic acid, the greater will be the number of entrained particles of spent acid. This is presumably due to both the reduced residence time and the fact that the turbulence zone increases in height with increasing flow rate. These facts determine, then, the maximum acceptable rate of flow for a given settling vessel and desired degree of separation.

Approach to equilibria = Seat acid concentration at given velocity × 100

3.6 EFFECT OF TOLUENE SULFONIC ACID

The composition of the resulting phases depends not only on the efficiency of separation but also on the amount of toluene sulfonic acid in the acid mix. With no toluene sulfonic acid, the concentrated acid mix from a continuous system generally contains about 10.4% spent acid, 88% AB sulfonic acid plus all of the unreacted alkyl benzene. This corresponds to a ratio of 8.5:1 of ABSO to 78% HSO. If toluene sulfonic acid is present, it solubilizes some of the spent acid into the sulfonic acid such that, even at equilibrium, the concentrated acid mix contains more spent acid. The more toluene sulfonic acid present, the lower will be the ratio of ABSO:78% HSO.

The composition of the spent acid layer is generally unaffected by the presence of toluene sulfonic acid. At equilibrium, the spent acid layer contains practically no sulfonic acid. However, in factory systems, spent acid normally contains about 1–2% sulfonic acid. Furthermore, many of the side reaction products of the sulfonation dissolve in the spent acid contributing to its dark red color.

3.7 PASTE COMPOSITION

Most finished product formulations require a somewhat greater percentage of sodium sulfate than that normally made from the spent acid remaining in the concentrated acid mix. This additional sodium sulfate can be supplied by:

(a) running two sulfonation systems, one to concentration, and one directly to neutralization.
(b) concentrating all of the acid mix and returning a portion of the spent acid to neutralization.
(c) adding sodium sulfate to the crutcher.

Method (a) is no longer commonly used. The choice between (b) and (c) must be a local one, based solely on the economics of making versus buying sodium sulfate.

3.8 OPERATING PROBLEMS

The goal of dilution and settling is to remove a constant quantity of spent sulfuric acid from the acid mix stream and to deliver to the neutralization equipment at a constant rate of flow, concentrated acid mix of uniform quality.

3.9 BATCH PROCESS

Dilution and settling follows aging of the sulfonated acid mix. The aged mix is recirculated through a dilution heat exchanger, while recycled spent acid corresponding to the amount required to change the acid mix phase system to a continuous sulfuric acid phase is gradually added to it. The recycled spent acid is normally added before the dilution water, since less heat will be evolved with the addition of a given quantity of spent acid than with the same quantity of water. Addition of all the recycle acid reduces the viscosity of the acid mix sufficiently to increase greatly the heat transfer efficiency in the dilution cooler. This, in effect, increases the cooling capacity of the heat exchanger so that it can more readily remove the heat evolved when the dilution water is subsequently added. This procedure thus places the minimum peak heat load on the cooler.

The rates of addition of recycle acid and of dilution water are adjusted such that the completely diluted acid mix is at the desired settling temperature. After settling, the spent acid is drawn from the settled mix, part being sent to storage for eventual disposal and part to the spent acid tank for recycle to the next sulfonation batch. The concentrated acid mix, together with the remaining spent acid needed to meet the formula requirements for sodium sulfate, is then sent to neutralization.

The strength of the spent acid can be checked by measuring its specific gravity. Published specific gravity data for sulfuric acid are sufficiently accurate for the batch process since the spent acid sample generally contains a negligible quantity of sulfonic acid.

3.10 CONTINUOUS PROCESS

In a continuous system, the heat of dilution is controlled in a conventional dominant bath such as that used for sulfonation. The acid mix is fed from the sulfonation system directly to the outlet of the cooler while the recycle acid and the dilution water are added at the dilution pump.

3.11 CONTINUOUS SETTLING SYSTEM

Dilution of acid mix to the specified spent acid strength is extremely important for control of both neutralized paste composition and for diluted acid mix corrosiveness. However, since water is relatively easy to meter, there is usually little difficulty in accomplishing this. Occasionally in continuous systems, fluctuations in flow rate will occur due to inadequate supply pressure or excessive back pressure from the dilution dominant bath. Perhaps one of the more common faults in operation is the tendency to over-correct dilution water usage when drifts in spent acid strength do occur. Specific gravity is again used to measure spent acid strength; however, since the spent acid contains some sulfonic acid, the specific gravity for a given spent acid strength is lower than would be obtained from published data for sulfuric acid.

Poor dilution temperature control will result in poor finished product color and in base odor development. It is commonly caused by the same problems which cause poor temperature control in sulfonation. Insufficient recycle of spent acid with a resultant reversion of the phase system to the more viscous continuous sulfonic acid phase can also cause poor temperature control. In order to avoid color and odor problems, temperatures must not exceed maximum permissible levels at any time during dilution or settling.

Poor settling is caused most commonly by insufficient spent acid recycle or from air in the acid mix. The air can originate in the sulfonation system or from packing gland leakage in the dilution dominant bath. This can be prevented by proper control of back pressure to insure against cavitation of the dilution pump. As in sulfonation, poor back pressure control can also result in poor recirculation and localized high temperatures.

Metering acid mix is extremely difficult. Since good control in neutralization is very dependent upon a uniform flow of this material, other steps must be taken to ensure uniformity. Since all of the stocks to sulfonation and dilution are independently metered, it is assumed that the flow of diluted acid mix to the settling tank is constant. By carefully metering the spent acid removed from the system, the flow of concentrated acid mix is controlled by difference. The position of the interface in the settling tank is checked frequently to ensure that the proper quantity of spent acid is being removed from the system. If the interface drops below the desired position, the rate of spent acid draw-off to storage is decreased and vice versa. There is difficulty in seeing the sulfonic-sulfuric acid interface in the sight glass and also some doubt that this interface truly reflects that in the settling tank. When spent acid is returned to neutralization, careful control of flow rate must be exercised for good paste and composition control. Returning too much spent acid will result in an excess of sodium sulfate and will, in effect, dilute the finished product. For shutdowns of less than 96 hours, common practice is to leave the settling tank full. For longer shutdowns the settling tank must be emptied.

4 Neutralization

The object of the neutralization step is to neutralize the acid mix and produce a uniform paste of the proper composition at a rate sufficient to supply the crutching operation. The concentration, moisture content, temperature, and density of the neutralized paste must be carefully controlled. One of two materials is commonly used for acid mix neutralization, sodium hydroxide or sodium carbonate.
Use of Sodium Hydroxide:

4.1 CHEMISTRY OF NEUTRALIZATION

The neutralization of acid mix is relatively simple:

These are the highly exothermic reactions of strong acids with strong base:

$$H + OH \rightarrow H_2O + 13.6 \text{ k-calories (dilute solution)}$$

The heat of neutralization of the actives can be taken as approximately 13.6 k-calories per mole of water formed. The heat of neutralization of HSO depends on the spent acid strength, the lye concentration, but is usually in the range of 2–31 k-calories per mole of water formed. The excess over 13.6 k-calories is due to the heat of dilution of the spent acid and caustic soda. Of the total heat of neutralization of unsettled ABS acid mix (from oleum sulfonations) about 65–75% is due to the neutralization of spent acid.

The major variables which must be controlled during neutralization are:

(A) Paste pH

The pH of finished product is well buffered at approximately 110.0 by both the phosphates and the silicate. As a result, the pH of the granules is not often out of limits. Nevertheless, variations in paste pH, even if not reflected in the finished product, can greatly affect product quality. Thus:

(i) If at any time the paste pH drops below 7, hydrolysis of alkyl sulfates will occur.
(ii) Finished product and paste colors of both ABS and AS formulations tend to suffer if the paste pH ever drops below about 7. This is believed to be

DOI: 10.1201/9781003596219-5

due to the dissolution of metal oxide films which build up under alkaline conditions within the lines and vessels of the paste system. This film protects the equipment against further corrosion. Acid paste will dissolve the film, forming soluble metal salts, e.g., FeO, which reprecipitate under alkaline conditions as $Fe(OH)_5$ or FeO_3, and color the paste.

(iii) Precipitation of solid SiO (floc and grit) increases at lower finished product pH. Even if the product pH is within limits, however, silica precipitation tends to increase if the paste pH is low since the acid takes NaO from the silicate. By this mechanism, silicate acts as a buffer against low crutcher mix or product pH.

(iv) Base odor development on aging of the finished product apparently tends to increase when paste pH is low. As in the case of floc, this will be true even if the finished product pH is within limits. The mechanism involved is not fully understood but is believed to involve the formation of acidic odor bodies.

(v) Detergency is optimum at about pH 10.

(vi) Mildness of the finished granules decreases with increasing pH. This will, of course, be of importance primarily in hand washing applications.

(b) Paste Moisture Content

Moisture content is selected to give the proper amount of water for crutching and spray drying. A certain minimum amount of water (45–50%) is necessary to keep the neutralized paste fluid enough for easy handling and good heat transfer. In high active formulations, this limits the minimum attainable crutcher mix moisture. Where paste is weighed into the crutcher, uncompensated variations of paste moisture will result in variations of finished product composition. In batch plants where the neutralizer is also the crutcher, variations of paste moisture will not affect finished product composition but will cause variations in granule density, particle size, etc.

(c) Caustic Strength

In order to maintain uniform paste pH and paste moisture content, the strength of the caustic solution used must remain constant. Material balances and formula cards generally specify the caustic soda as NaOH and show the total water required in neutralization; both are added as caustic solution and the process water, as a single quantity. Each factory must then decide on a caustic strength that is most suitable for its operation and calculate the additional quantity of process water required.
Selection of the desired caustic strength normally is based on:

- Local availability of caustic solution.
- Desired strength for other processes in the factory.
- Limitations on storage capacity.
- Expected minimum storage temperature.

Control of crutcher mix temperature is important. In addition, however, temperature has a large effect on paste viscosity. Thus, while increasing temperature generally reduces viscosity, with ABS paste a phase change occurs which causes the viscosity to increase to a maximum. The temperature at which this phase change occurs depends, to a large extent, on the paste moisture and sulfate contents. Normal practice is to use the temperature which gives optimum crutching conditions, provided only that the paste is fluid enough to handle.

A measure of the degree of aeration of a paste is used. Aeration is not normally a serious problem with the products made in most of factories although it can be with high density formulas. Extreme aeration is bad in all cases because it hurts both recirculation within the paste system and also subsequent processing.

Highly aerated pastes are viscous and have a whipped cream appearance. The dilution pump is used to boost acid pressure and a rotameter is used for control of flow rate. The volume flow of spent acid necessary to give the proper flow of concentrated acid mix can be calculated from the specific gravity relationships and the material balance. The important thing is that process water usage be matched to acid mix and caustic flow rates in order to maintain proper paste composition. The time required to empty the settling tank can be reduced by gradually raising the interface in a controlled manner during the last day of normal operation.

The spent acid which is removed from the system contains up to 1% organic material, including some alkyl benzene sulfonic acid. (Thus far it has not been considered worth recovering the sulfonic acid present in the spent sulfuric acid.) This spent acid is entirely suitable for many industrial uses such as superphosphate fertilizer manufacture and steel pickling. In some countries where both the super active (35% ABS) and lower active products are made, substantial savings are realized by reducing oleum usage on the lower active formula and making up the required sodium sulfate by returning spent acid from storage.

Acid mix can be neutralized in either a batch or a continuous system. In a batch neutralization process, all the process water and caustic required for the neutralization of one batch of acid mix are first added to a tank which is equipped with an effective agitator. The total NaOH required is determined by the amount and composition of the acid mix. The total amount of water required is determined by the final crutcher mix moisture desired.

Acid mix is added to caustic slowly enough that the pH is always kept on the alkaline side. As the diluted caustic solution is circulated through the external heat exchanger, acid mix is added at the recirculation pump. The cooling water on the heat exchanger is adjusted to maintain the specified paste temperature. Good mixing is essential because, when partially neutralized, acid mix forms a gummy phase which tends to prevent complete neutralization. This condition is readily detected because the gummy phase has a purplish discoloration and appears as streaks. On the other hand, excessive mixing can result in considerable air entrainment. Thus, initially during neutralization, while the viscosity of the mix is still low, adequate mixing is obtained within the pump alone and the agitator in the tank is not used. Toward the end of the neutralization, the agitator must be turned on periodically. It is generally used as little as possible, consistent with obtaining good mixing and minimum air entrainment (low paste density).

The paste pH is checked manually with pH indicator solution. A small amount of the formula usage of caustic is withheld so that the final paste pH can be adjusted to the desired value.

4.2 CONTINUOUS NEUTRALIZATION

A dominant bath system, fitted with or without a surge tank, is well suited to continuous acid mix neutralization. Caustic, process water, and acid mix are added into the recirculating paste stream. Many variations of the dominant bath loop are in use. Caustic and water may be added at the suction of the recirculation pump or in a separate caustic-mix-pump; acid mix may be added at a separate mechanical neutralizer or at the periphery of the recirculation pump. As in the batch process, paste pH must be kept on the high side and good mixing is essential because acid mix forms a gummy phase when partially neutralized. The problems of process control are somewhat more difficult than in the batch process.

4.2.1 pH Control

For proper pH control, the instantaneous pH must be properly sensed and the correct signal must then be applied to the caustic control valve. A representative sample of the neutralized paste stream must be constantly delivered to the pH electrodes.

For most rapid response, the sample should be taken from some point in the neutralization loop after the addition of caustic and acid mix but before the surge tank. If this is a point of low pressure, a small gear pump can be used to deliver paste to the pH electrodes. Some of the older factories are not equipped with this gear pump and therefore take the paste sample at the discharge of the recirculation pump. With thin paste and high recirculation, these plants maintain satisfactory pH control. However, with viscous paste (low recirculation), the increased time lag due to the surge tank may cause unstable pH control. Under these conditions, the sample is not representative of that leaving the point of neutralization. Both the gear pump and the long-time lag can be eliminated by feeding both caustic and acid mix at the paste recirculation pump and taking the sample before the heat exchanger.

Constant feed rates and compositions of acid mix and caustic are necessary for good pH control. Pressure upsets in the neutralization loop cause feed rates to vary, and such upsets should be minimized. If the paste recirculating pump is used to transfer paste to the crutcher, care must be taken to prevent large pressure fluctuations. Most plants are now equipped with a separate paste transfer pump to circumvent this difficulty.

Once a representative paste sample is obtained and delivered to the electrodes, the chief problem remaining is to sense correctly the pH and to convert this signal into a movement of the caustic control valve. Proper sensing requires properly selected, clean electrodes. Manufacturing standards specify the minimum frequency for washing the electrodes. The pH signal which is generated is transmitted to a pH controller. Here the electrical signal is changed to an air signal, which moves the caustic control valve.

Each neutralization system has its own combination of reset, proportional band, and rate settings for the controller, which will give optimum pH control. This combination of variables differs not only from system to system, but with different formulas made on the same system. When determining optimum settings, sufficient time should be allowed for the system to come to steady state before making further changes.

4.2.2 Control of Paste Moisture Content

The principal variables affecting moisture content are caustic soda strength and process water metering. Any caustic strength is satisfactory providing the solution does not crystallize, cause extreme corrosion, or exceed the metering capacity.

However, once a decision on caustic strength is made, strict adherence to this level is of utmost importance. Small, desired changes in paste moisture content can then readily be made by varying the process water rate. It is normally considered better to increase paste moisture content when using large quantities of dry scrap than to add water to the crutcher. Paste moisture is checked frequently by analysis. In addition, the experience of the operator in visually gauging paste fluidity will be of great value.

One uncommon but potential case of variation in paste moisture content is by water leakage into the system, either through a leaking heat exchanger, washout line, or pump packing gland.

4.2.3 Paste Temperature

For optimum crutching conditions, the best paste temperature must be determined experimentally and then adhered to as closely as possible. Normally, this is easy to do. However, the efficiency of heat transfer in the paste heat exchanger can become reduced as a result of several causes, any of which may result in inability to hold down neutralization temperature:

Insufficient cooling capacity. In some factories, the cooling water temperature is too high during summer months to provide adequate paste cooling. Inadequate cooling can also be due to a reduction in cooling water flow because of fouling of the water side of the heat exchanger, problems with the cooling tower, or excessive demands by other processing units. Gradual fouling of the heat exchanger is almost inevitable in areas where the mineral content of the cooling water is high. It must be accepted and time allowed for adequate cleanouts.

Make-up on the stock side of the heat exchanger is not too common with ABS paste when daily steam outs are made. However, very cold water can cause pastes containing fatty alcohol sulfates to solidify on the surfaces of the tubes so that the heat transfer is reduced even though the average temperature difference between the paste and water is increased. In the extreme, the exchanger can become completely plugged. These conditions, which can be corrected by warming up the heat exchanger, determine the minimum safe cooling water temperature.

If the paste viscosity becomes too high, it will cause both poor heat transfer and a low recirculation rate. A large spread between the paste temperature in and out of the heat exchanger is the best indication of reduced flow in the circulating system, whether it is due to fouling of the tubes or to high paste viscosity. Reduced paste recirculation rate can be due to cavitation of the paste circulating pump.

4.2.4 Control of Paste Density

Paste density can readily be determined by filling a vessel of known volume with the paste, weighing it and calculating the absolute density. This should normally run

0.9–1.1 g/cc, preferably toward the higher end of this range. If unduly low paste densities are found, the system should be checked for possible sources of air. The most likely source is the paste circulating pump packing. Inadequate lubrication or cavitation will often be observed.

4.3 USE OF SODIUM CARBONATE

In some countries, sodium carbonate (soda ash) provides a cheaper means of neutralizing acid mix than does sodium hydroxide:

2 RO–SOH + NaO ->2 R–O–SON + H,O + CO

These are reactions of a strong acid with a weak base and can be simplified:

2H + CO > O + CO + 2,3 -calories (dilute solutions)

Thus, the heat of neutralization with carbonate is appreciably less than with caustic soda. In some plants, heat exchangers in the paste systems have been eliminated and paste temperature is controlled by controlling carbonate solution temperature. This normally gives temperatures of 80–90 °C, which are often satisfactory with the low-active formulas. If colder paste is required (as with the higher active formulas), an external paste heat exchanger is provided.

Most of the comments made on caustic neutralization are equally applicable where carbonate is used. The only major new problem arises when sodium carbonate is used for CO removal. Incomplete removal of CO results in poorer behavior in both the paste system and in the tower. Excessive retention of CO in the neutralized paste hurts paste circulation and causes frothing in the surge tank, resulting in loss of control on paste temperature and pH. It also increases crutcher mix viscosity and causes poorer spraying, giving product of poor granule structure, wider spread in particle size and erratic density control.

Completeness of CO evolution from paste depends on:

4.3.1 Neutralized Paste pH

CO retention increases with increasing pH. When the paste pH is below 6, CO evolution, at equilibrium, is essentially complete. For pastes that are slower or more difficult to degasify, however, a partially neutralized paste pH of about 3.0 is optimum. Consequently, carbonate neutralization is a two-stage operation. Some 95% to 98% of the NaO required is first furnished by sodium carbonate and the neutralization is then completed using sodium hydroxide in the conventional manner.

Control of the final paste pH is not greatly different from caustic neutralization. More variables are present in the system, however, so even stronger efforts must be made to secure the right combination of operating conditions. In most systems, the partially neutralized paste stream and a recirculated paste stream containing excess caustic are discharged to the top of the paste in the surge tank and the pH sample taken at the bottom of the tank. This creates an inherently long process time lag in the system.

In addition, good mixing in the surge tank is essential. The pH from the de-gasifier is normally controlled in the range 2–5 by varying the amount of sodium carbonate which is metered into the system. This pH is checked manually every few minutes with a suitable pH indicator. Extreme fluctuations of the pH of the partially neutralized paste occur only if fluctuations in acid mix flow rate or composition occur. Sometimes strainer make-up or pumping problems in the carbonate solution system can cause pH fluctuations.

Some of the newer systems use the caustic and paste water as added in a small in-line mixer before the surge tank and the pH is sensed at the discharge of the mixer. This set-up eliminates the long-time lag and the intense agitation needed in the surge tank, but makes it more difficult to correct for momentarily high or low pH.

4.3.2 Paste Formulation

CO retention tends to increase with increased active content of the paste. This is attributed to increased paste viscosity and can be compensated for, in part, by putting more water in with the carbonate solution (lower strength) and reducing process water usage. Due to the large quantity of water which may be required for the sodium carbonate solution, very little, if any, water is left for direct injection into the process system. With high active paste from concentrated acid mix, all of the water should go in with the carbonate solution to ensure adequate mixing and CO evolution. This procedure also ensures more uniform paste moistures. Changing paste moisture content is difficult, however, due to the large mass of sodium carbonate solution whose strength must be changed prior to a paste moisture change, but this is not a serious defect.

4.3.3 Acid Mix-Carbonate Mixing

Efficient mixing of acid mix and carbonate solution is necessary for uniform pH control and complete CO evolution. Mixing is achieved by passing the acid mix and sodium carbonate streams through a series of orifice plates in the carbonate neutralizer column. The size of these orifices should be as small as possible, consistent with the pressure available. A margin of safety is provided in sizing orifices due to the gradual buildup which occurs during operation. Periodic steam out of the column, however, is required. Special modifications to the columns are required to obtain the necessary intense mixing needed for making mixed active pastes. The pastes obtained from caustic or carbonate neutralization behave essentially the same. The choice between the two methods is based solely on the economics at each factory.

4.4 CAUSTIC SODA AND SODIUM CARBONATE DISSOLVING

As has been noted above, close control of caustic soda and sodium carbonate solution strengths is necessary. The use of specific gravity determinations and suitable charts of strength versus specific gravity and temperature are adequate; however, titrimetric checks in the laboratory are recommended.

Caustic soda may be purchased as a solution. Often, however, solid caustic soda is used and it must be dissolved at the factory. The principal problem is the high temperature generated due to heat of solution. Normally this temperature rise can be tolerated. Sodium carbonate is received in powder form and must be dissolved. Excessive sludge or impurities must be eliminated by adequate settling or straining.

4.5 PHASE SEPARATION

When neutralized paste is allowed to stand at 45 °C or higher, a layer of aqueous sodium sulfate solution will settle out. This effect can happen in the surge tank during an over-night shut-down and it is, therefore, important that the first paste to crutching be well agitated to insure uniformity.

This effect is sometimes used to advantage in the laboratory to concentrate both ABS and alkyl sulfate pastes. Active:sulfate ratios can often be achieved. The process, however, is quite slow (hours) and difficult to control because of the large differences in degree and rate of settling which occur with small changes in operating variables. It is therefore never used in the factory.

5 Paste and Crutcher Mix Phase Systems

5.1 DEFINITIONS

A system is a substance or mixture of substances isolated for study of the effects of temperature, pressure, and change in concentration on the state of materials present.

A phase is a homogeneous, physically distinct, and mechanically separable (e.g., by filtration, settling, and hand picking of crystals) portion of a system.

The number of components of a system is the smallest number of independently variable constituents needed to specify the composition of each phase of a system.

ABS paste is composed of three components (neglecting un-sulfonated AB), namely ABS, NaSO, and water. A mixture of these components will separate into one or more phases such as lye, crystalline ABS, or crystalline Na_2SO.

The number of phases present, the composition of each phase, and the relative amounts of each phase depend on the overall composition of the mixture and also on the temperature and pressure of the system. Assuming the pressure and temperature are such that an equilibrium involving a vapor phase cannot exist, then the maximum number of liquid or solid phases that can co-exist in a three-component system is three.

5.2 THE PHASE DIAGRAM

For a given temperature and pressure, the various phases that can exist can be represented graphically on an equilateral triangular diagram. The apexes of the triangle represent the pure components (100%) while the base opposite represents 0% of that component. The percentage of any component is determined by drawing a line perpendicular to the line bisecting the apex and noting the ratio of the distance from the point of intersection to the base as compared with the distance from the apex to the base.

A generalized diagram for a surfactant, electrolyte, water system describes the surfactant may be ABS, alkyl sulfate, soap, etc., and the electrolyte may be NaSO, NaCl, etc. The solid lines or curves within the triangle represent the appearance (or disappearance) of certain phases, as the overall composition is changed. The areas

DOI: 10.1201/9781003596219-6

containing tie lines represent systems of two phases and the composition of each phase is represented by the points at the ends of the tie lines.

5.3 POSSIBLE PHASES IN SURFACTANT-ELECTROLYTE-WATER SYSTEM

5.3.1 Isotropic Liquid Phases

These are water-rich solutions whose optical properties do not depend on the axis along which observation is made. Any composition within the boundary has only one isotropic phase. If, however, the overall composition has two isotropic phases present, then these are:

Lye phase: Generally, lye phases contain very little e.g., in the system ABS-NaSO-water a lye phase < 1% ABS, 10–30% NaSO, and 70–90% water. Other phases generally contain more surfactant and less electrolyte than does the lye phase. Thus, the ABS phase contains 3–35% ABS, 0–10% NaSO, and 65–95% water. Within the area, the term isotropic phase is preferable to either the term lye or ABS since there is only one phase present.

5.3.2 Anisotropic Liquid Phases

These are surfactant rich phases whose optical properties depend on the axis along which an observation is made. (That is, they rotate the plane of polarized light.) These phases are called "liquid crystalline" since they behave as viscous liquids but also have optical properties that are characteristic of crystalline materials. Two such phases exist:

(i) Neat phase: This phase is represented by the area characterized by having high surfactant and low electrolyte contents; e.g., in the ABS system a neat phase contains more than 70% ABS, very little electrolyte, and 25–30% water.
(ii) Middle phase: This phase (M) contains less surfactant and more water than neat. It occurs in tallow alkyl sulfate and soap systems but not in ABS systems except at low temperatures. The system changing temperature gradually shifts the phase boundaries such that at certain temperatures a phase may disappear entirely.

5.3.3 True Crystalline Phases

These are solid phases having all the properties of an ordinary crystalline material.

(i) Crystalline surfactant phase: This phase is found in soap and alkyl sulfate systems but not normally in ABS systems due to the numerous side chain possibilities for the AB. The phase is essentially all surfactant but may contain small quantities of electrolyte and water.

(ii) Solid electrolyte phases: These phases contain no surfactant but may contain hydrated water.

5.4 MULTI-PHASE SYSTEM

The phases discussed above can exist either alone or in combinations depending on the overall composition and the temperature (and pressure). The possible combinations are:

(A) One-Phase System

These are systems whose overall composition lie in the areas designated neat. M (middle) or I (isotropic), which have been discussed above.

(B) Two-Phase Systems

These are systems whose overall composition lies in an area containing tie lines. If the overall composition varies along a tie line, the composition of each phase remains constant (ends of the tie-line) but the amount of each phase changes.

(C) Three-Phase Systems

These areas represent overall compositions which would appear as three phases. The composition of each phase is fixed and is represented by the apexes of each triangle. Variations of overall composition within the area changes the relative amounts of each phase present but not their compositions.

The various boundary lines (or contours) depend on the nature of the surfactant and electrolyte. In general, replacing NaSO with NaCl or NaOH as well as lowering temperature causes the boundaries to shift to the left. Presence of STS (in ABS systems) and use of more soluble ABS shifts the boundaries.

In any given system, the phases may settle out or may form emulsions or suspensions. When emulsions do occur, systems in which the continuous phase is neat or middle tend to be viscous. If the continuous phase is a lye or weak phase, the resulting mix tends to be much thinner. One of the major deficiencies of the phase approach is that it does not indicate which phase is the continuous one.

5.5 LOW ACTIVE ABS PASTE

By low active is meant pastes that result from the direct neutralization of acid mix sulfonated with sulfuric acid or oleum. The composition is roughly 25–30% ABS, 20–25% NaSO, and 45–55% water. In this range of compositions, the system is either in the three-phase system of neat, lye, and crystalline NaSO, or the two-phase system of neat and lye. At temperatures above 50 °C, the mixture consists of a grainy suspension of neat-in-lye which separates readily on standing. If crystalline NaSO is present, it remains emulsified in the neat phase. Above about 55 °C, however, the paste phase system changes to an emulsion of lye-in-neat. This emulsion is so finely grained that the droplets of lye cannot be 800 X. Only after seen under a microscope even at a magnification of several hours in an ultra-centrifuge will a small amount of lye phase separate. Interestingly, as little as 8% neat phase is able to emulsify up to 92% lye phase and the electrical conductivity of the emulsion obtained, in which the neat is the continuous phase, will not be much higher than that of pure neat. The tendency to form a lye-in-neat emulsion in an ABS paste increases with increasing temperature and is strongest at 70–110 °C. However, emulsion formation is sensitive to the type of inorganic salt present. Sodium silicate, and monovalent salts such as sodium chloride, prevent the formation of an emulsion at all temperatures up to 110 °C. Most polyvalent salts such as NaSO, NaCO, and phosphates favor emulsion formation at temperatures above 55 °C. A few polyvalent salts (sodium citrate) tend to reduce emulsion stability.

5.6 HIGH ACTIVE ABS PASTE

These are the pastes derived from either settled acid mix or from SO sulfonation. Paste from settled acid mix generally forms either a neat-nigre-lye or a neat-nigre system. If there is lye present or if the nigre is weak (low ABS), emulsions with neat will still form. If the nigre is strong, the emulsions are normally unstable. Thus, paste from SO sulfonation forms a neat-strong nigre system which settles out rapidly. If the isotropic phase has a yellow color and becomes turbid on cooling as a result of separation of neat, we call it a nigre phase.

5.7 ABS CRUTCHER MIX SYSTEMS

The main ingredients present in crutcher mixes of heavy-duty formulations are ABS, sodium sulfate, sodium tripolyphosphate, sodium silicate, and water. The phase system of the paste itself, particularly if it is a fluid nigre paste, will normally be changed radically in the crutcher from what it was in the paste. The same phases are probably encountered in crutcher mixes as in paste: neat, lye, nigre, solid inorganic and crystalline organic phases. When a sample of crutcher mix is centrifuged, three layers generally are obtained:

1) The top layer consists mainly of neat phase. Depending on the centrifugal force used, this layer may include more or less emulsified lye phase and entrained solid electrolyte.
2) The middle layer may consist of either lye or nigre phase. This phase is normally lye unless alcohol is added or extra-large amounts of STS are used. Nigres seldom are seen in ABS crutcher mixes.
3) The lower layer consists mainly of solid electrolyte crystals with entrained lye and surfactant phases.

Experiments show that the lye phase contains STPP in solution, at equilibrium. The STPP used usually consists of about 25% form I and 75% form II. It is generally accepted that STPP form I will hydrate immediately on contact with water (or lye phase) to form a hexahydrate. However, the formation of hexahydrate from STPP form II involves a two-step mechanism in which first the STPP form II dissolves in the water and then the hexahydrate crystallizes from the STPP form II solution. Therefore, in the initial stages of hydration, the lye phase is super saturated with STPP.

The solid electrolyte fraction contains those salts (mainly anhydrous or hydrated STPP and NaSO) which are present in the crutcher mix in quantities greater than the amounts which can dissolve in the lye phase. The solubility of silicate is so high that it is never present in the solid phase other than in the liquid entrained by the inorganic crystals or adsorbed on the solids.

Most of the water that is removed from the crutcher mix during spray drying comes from the isotropic (normally lye) phase. The lye phase products (principally inorganic salts) which precipitate as the water is driven off tend to coat the surfaces of the finished granules.

Most of the surfactant is beneath the surface of the granule. Thus, alcohol extraction of granules to remove surfactant does not normally affect the granule structure. Under certain processing conditions, however, some surfactant may also be present on the granule surface.

5.8 PASTE AND CRUTCHER MIX PHASE SYSTEMS

For example, granules blown from a nigre phase (such as may be produced by addition of alcohol to crutcher mix) will be found to contain, as part of the surface of the granule, much of the ABS from the nigre. Such granules are usually very soft and sticky and will break down on extraction with alcohol.

Nigre in a crutcher mix can be reduced or eliminated by replacing sodium sulfate with sodium chloride, reducing or eliminating STS, lowering temperature, and minimizing free moisture in the crutcher by maximizing phosphate hydration or lowering crutcher moisture.

5.9 EFFECTS OF SODIUM TOLUENE SULFONATE IN ABS SYSTEMS

Since STS is such an important ingredient in many formulas, its effects are summarized here:

1) STS solubilizes ABS. This is equivalent to shifting the phase boundaries to the right, that is, enlarging areas of nigre formation. It is in the direction of increased stickiness.
2) STS induces more ABS crystallization. This may provide decreased granule stickiness and better caking resistance. STS decreases paste and crutcher mix viscosities. This is probably due to a change in the type of emulsion formed (neat-in-lye instead of lye-in-neat). Improved atomization of crutcher mix results.

Whether 1 or 2 will predominate depends to a large degree on the formulation. Thus, with formulas containing less than about 35% ABS and more than about 10% NaSO, the increased ABS crystallization effect will predominate. However, formulas containing more than 35% ABS and less than about 10% NaSO may actually become stickier due to the solubilization of ABS in a nigre phase.

5.10 ALKYL SULFATE PASTE AND CRUTCHER MIX SYSTEMS

Alkyl sulfates paste and crutcher mix phase systems are much more similar to those of soap than to ABS. They differ from the corresponding ABS phase systems primarily in that:

1) Alkyl sulfates have a strong tendency to form a nigre phase instead of a lye. Neat and nigre phases form emulsions, and this is one reason, presumably, why alkyl sulfate pastes and crutcher mixes are much thinner than those of ABS with the same moisture content.
2) Alkyl sulfates crystallize fairy readily. The AS in finished product is crystalline regardless of the phases system from which it is blown. ABS granules on the other hand are not in the form of solid crystals. For this reason, AS granules have less tendency to be sticky than ABS granules, and also as can be seen from equilibrium moisture data, the AS in finished product is less hygroscopic than is ABS.
3) Alkyl sulfates can form a middle phase which ABS does not.

6 Crutching

Crutching is a complex mixing operation. In the crutching step, all the ingredients that constitute the base granules are brought together. The surfactant, in the form of neutralized paste, large quantities of dry builder, and a variety of other ingredients, both solid and liquid, must be thoroughly mixed in a limited time. The resultant slurry, uniform in composition and free of lumps, is then spray dried to produce finished base granules.

The essentials of an effective crutching system are:

(a) Proper formula through precise scaling or measuring.
(b) Good mixing of all ingredients.
(c) The correct level of STPP hydration.
(d) Proper characteristics for spray drying.
(e) Uniform operation with minimum batch-to-batch variation.

Failure to meet the above requirements will result in erratic operation, reduced rates, density fluctuation, excessive scrap, non-homogeneous product, or excessive equipment makeup problems.

Two most important characteristics of a crutcher mix, atomizability and puffability, will first be defined. Both of these properties are closely related to tripolyphosphate hydration which will, therefore, be discussed in some detail. The effects of other constituents such as toluene sulfonate, salt, nonionics, and even air will also be discussed. In addition, the discussion will cover a description of crutching equipment and methods of operation.

6.1 ATOMIZABILITY AND PUFFABILITY

6.1.1 Atomizability

Atomizability is the ease with which a crutcher mix is broken up into droplets at the spray nozzle or disk. It is determined largely by the viscosity and surface tension of the crutcher mix and can, therefore, be altered by changing the composition or

 DOI: 10.1201/9781003596219-7

temperature of the mix. It cannot, however, be altered by changing, for example, the type of nozzle or the spraying pressure used.

Thus, increasing the spraying pressure improves atomization but does not affect atomizability. Improved atomizability results in reduced particle size at the spray nozzle. Any change which reduces viscosity or surface tension should improve atomizability.

6.1.2 Puffability

Puffability is related to the tendency of the atomized droplets to undergo a volume increase during the drying step. It is determined largely by the air and free moisture contents of the crutcher mix and probably also due to the type and stability of the emulsion formed. Like atomizability, it is a property of the mix. It is unaffected by, for example, increasing the air temperature in the tower, which increases puffing but not puffability.

6.1.2.1 Air Puffing

During crutching, a considerable amount of air is entrained into the mix. As the crutcher mix is pumped to the nozzle, it is compressed and much of the entrained air disperses into the mix. As the crutcher mix leaves the nozzle, this dispersed air is released by the reduction in pressure. This reduction of pressure and also the increase of particle temperature causes the granule to puff. In extreme cases, complete fragmentation of the droplets can occur.

Air puffing is a significant fraction of the total puffing when pressure atomization is used. With spinning disk atomization, the much lower pressure drop through the disk probably reduces the contribution of air puffing to total puffing.

6.1.2.2 Steam Puffing

Steam puffing results when moisture in the interior of a particle vaporizes faster than it can diffuse to the surface. Under these conditions, the particle expands until the drop surface becomes hard and will not yield further, or until the steam finds an outlet to the particle surface. Only free moisture (un-hydrated water) is effective in causing steam puffing. Air and steam puffing are quite different from one another. The large center void results from steam puffing. The small voids in the cellular wall are from air puffing.

6.1.3 Crutcher Mix Phase System

Atomizability and puffability are closely related to the phase system of crutcher mix. The most common liquid phases present in crutcher mix are neat and lye. The relative amounts and distribution of these phases have a significant effect on atomizability and puffability. Thus, for example, increasing the quantity of lye phase in a mix decreases crutcher mix viscosity and thus improves atomizability. Furthermore, if lye is present as the continuous phase, the viscosity is lower than if it is emulsified in the neat phase. Puffability is also affected since the amount of air that a mix can entrain is related to the viscosity of a mix. Also the rates of diffusion of moisture from the interior of a granule depend on the type of phase system present thus affecting steam puffing.

Although nigre phases are not normally encountered in crutcher mix, when they do occur, the mixes are difficult to atomize and the granules are sticky due to the presence of large quantities of surfactant on the surface of the granules.

6.1.4 Free Moisture of Crutcher Mix

Increasing the free moisture generally improves atomizability. Merely increasing the total crutcher moisture level, however, does not always increase atomizability since an accompanying increase in hydration may tie up the added water. Furthermore, the finished product particle size is not necessarily smaller since the wetter mix may increase particle agglomeration. This is discussed further under Spray Drying.

Increasing free moisture also increases steam puffing. For very light density products, increasing crutcher moisture is often the only means available for attaining the required density.

6.2 SODIUM TRIPOLYPHOSPHATE HYDRATION

Hydration of STPP is the only chemical reaction that occurs during crutching. A knowledge of its mechanism, of its effect on crutcher mix characteristics, and of means for its control is of prime importance in understanding the crutching operation.

Sodium tripolyphosphate reacts with water to form a hexahydrate. Hexahydrate in the finished product is desirable since, by tying up the water present in the granules, it gives a crisper product for a given finished product moisture level. The formation of hexahydrate also reduces the free moisture content of crutcher mix, thereby affecting atomizability and puffability.

6.2.1 Mechanism of Hydration

The mechanism of hydration of STPP in crutcher mix is thought to be:

Solid, anhydrous STPP first dissolves in water. The solubility relationships are unusual in that the hexahydrate is significantly less soluble than the anhydrous salt. The concentration of STPP in solution increases until the solution becomes supersaturated with respect to the hexahydrate, and at this time the hexahydrate begins to precipitate. As long as some anhydrous STPP remains, the concentration of STPP in solution will remain supersaturated and hydration will continue. If the crutcher mix is held for a sufficiently long period, all the anhydrous STPP will dissolve, the concentration of STPP in solution will decrease to the equilibrium solubility of the hexahydrate, and hydration will be complete. STPP hydration is strongly exothermic. The heats of hydration have been determined to be:

Form I = 50 cal/g Form II = 46 cal/g

These two crystalline forms, although chemically identical, differ markedly in their hydration behavior. Both, however, form the same hydrate and ion in solution. The rate of hydration of STPP depends on its solubility and its rate of solution. Form I is more soluble and dissolves more rapidly than Form II and, therefore, hydrates much more rapidly. The hydration of Form I is so rapid that for practical purposes, it is independent of crutching conditions. Form I is thus completely hydrated in the crutcher. The rate limiting step in hydration is usually the dissolution of STPP II. The overall rate and degree of hydration will be influenced by any factor that affects the rate of solution or the solubility of STPP II. Thus, strong electrolytes such as silicate reduce the solubility of STPP II and thereby retard hydration.

6.2.2 Variables Affecting Hydration

The rate of hydration of tripolyphosphate is a function of three major variables: the nature of the STPP itself, the composition of the crutcher mix, and the operating conditions.

6.2.2.1 Composition and Physical Nature of the STPP

(i) STPP I Content. The relative amounts of Form I and II in tripolyphosphate have an important effect on rate of hydration. Commercial tri-poly contains anywhere from 0% to 40% STPP II and most suppliers will furnish to specifications. Below 65 °C, the Form I content has little effect on rate of hydration since both Form I and Form II hydrate very rapidly. Above 65 °C, however, the rate of STPP II hydration slows appreciably while the rate of STPP I hydration remains very rapid up to about 95 °C.

(ii) Particle size can be an important variable, presumably because of its effect on rate of solution. Very fine particles hydrate very rapidly. Large particles or agglomerates which are not broken up in the crutcher may hydrate on the surface, forming a hard crust of hexahydrate which protects the interior of the particle from further contact with water and, thus, from further hydration. This is one potential source of grit in the finished product. It is particularly liable to occur if the conditions are such as to favor very rapid hydration or if the mixing in the crutcher is poor so that lumps of phosphate are not broken up. In practice, however, the effect of normal variations in particle size (about 50 microns average diameter) is small.

(iii) Purity of the STPP has a significant effect on its behavior. Pyrophosphates and orthophosphates reduce the rate of hydration of STPP by reducing its solubility. Glassy phosphates also reduce the rate of hydration. Moisture content of STPP is very important. Tripolyphosphate having a very low moisture content (about 0.05%) apparently contains relatively few nuclei or seeds for the crystallization of hexahydrate. The rate of hydration of the STPP II is reduced and also the hexahydrate crystals formed are larger in size and fewer in number. This is believed to be another potential cause of grit in the finished product.

The problem of low moisture content normally occurs only with very fresh tripolyphosphate. In most overseas countries, the STPP as received contains about 0.2% moisture and, therefore, grit from this source is not a problem. Too high STPP moisture content, on the other hand, is the most frequent cause of processing problems traceable to STPP. Increased moisture content increases the rate of hydration of STPP. Because increased moisture content in the majority of cases also means varying moisture content, the net result is a variation in the rate of hydration of the high phosphate which results in non-uniform processing.

In general, moisture levels above about 0.5% will give this type of problem. Also associated with STPP moisture contents exceeding 0.5% is an increased tendency to lump in the crutcher causing serious strainer make-up.

(iv) Moisture. Increasing crutcher moisture increases the rate and degree of hydration by increasing the rate of solution of STPP II. Higher crutcher mix moistures do not necessarily reduce the viscosity of a mix as much as would be expected since the added water may be tied up with the phosphates. Similarly, higher crutcher moistures do not necessarily increase the tower drying load since increased hydration permits blowing to a higher finished product moisture. High crutcher moistures may lead to problems such as grit formation, however, by causing excessively fast hydration. Lumps of STPP may hydrate on the surface before they have a chance to be broken up by the mixing in the crutcher.

(v) Nature of paste. STPP II hydrates at different rates in different pastes primarily because of variations in "availability" of the water. For example, the water in the lye phase of an emulsion is physically less available than that in free lye.

(vi) Electrolyte Content. Strong electrolytes, such as NaSO and silicate, repress hydration by reducing the solubility of STPP II.

6.2.2.2 Operating Conditions

(i) Time. Most crutcher mixes contain enough water to hydrate all the STPP if sufficient time is provided. The only control of crutcher mix aging time normally available, however, is obtained by varying the level of the mix in the drop tank. If the drop tank is maintained full, more time is available for hydration. If the level in the drop tank is maintained low, hold time is short and hydration reduced. Hold time is an unwieldy variable and of restricted value. However, its effect points up the need for smooth, regular operation in the crutching cycle.

(ii) Temperature. Crutcher temperature exerts a major control on rate of hydration of STPP II but has little effect on STPP I below about 95 °C. Hydration of STPP II proceeds very rapidly at about 50 °C. but as the temperature approaches 95

°C, hydration practically ceases. The inverse rate-temperature relationship is caused by the reduced solubility of anhydrous STPP II at higher temperatures.

(iii) Order of addition of crutcher mix ingredients affect STPP hydration. Adding the STPP early in the crutching cycle increases hydration since the moisture level in a partially completed crutcher can be quite a bit higher than the finished crutcher moisture.

6.2.3 STPP Reversion

STPP is degraded by water in two different reactions. These are discussed below.

(a) Crutcher Reversion

STPP hydrolyses in aqueous media, including crutcher mix, to form acidic ortho and pyro phosphates by reactions such as hydrolysis of STPP in crutcher mix is accelerated by higher moisture, higher temperature, and pH either above or below 10. Excessive hydrolysis results in the formation of sand or grit. It is not necessary for the overall pH to be lowered appreciably since the reaction can take place in localized areas of low pH. This is not a serious problem for most products. However, in products where very high processing temperatures are used, such as in very high-density brands, large amounts of grit can be formed, especially if the crutcher mix is held for longer than normal periods.

(b) Tower Reversion

When STPP hexahydrate loses water of hydration, for example, by over drying the granules in the tower, the STPP reacts with the water of hydration as it is driven off. The result is decomposition of the STPP and formation of more acidic ortho and pyro phosphates. In this case, the acidic phosphates react with the silicate in granules to form floc. Increased hydration in the crutcher increases the potential for reversion since un-hydrated STPP is not decomposed. The actual degree of reversion, however, is determined by the tower drying conditions.

6.3 EFFECT OF STPP HYDRATION

The degree of hydration of the tripolyphosphate in crutcher mix is important for several reasons:

(A) Crutcher Mix Viscosity

Water in the crutcher mix which is bound as STPP hexahydrate is removed from the lye phase and reduces the amount of this phase. This, in turn, concentrates the lye, precipitating solid components (STPP hexahydrate and NaSO). All these effects increase the viscosity of the crutcher mix and, therefore, reduce atomizability. At the same time, the reduced free moisture level of the crutcher mix reduces puffability.

The consistency of crutcher mix cannot be used as an estimate of STPP hydration since the size and shape of the hexahydrate crystals also affect consistency.

(B) Spray Drying

The degree of hydration must, therefore, be optimized. Too little and the product can become sticky. Too much and the result can be a thick mix with poor atomizability and excessive reversion. Sometimes the atomizability of a thick mix can be improved by raising the temperature of the mix after hydration has been completed. This can be done by injecting steam into the mix or by passing the mix through a heat exchanger on its way to the nozzle. Steam is used for temperatures up to about 105 °C; for higher temperatures, the heat exchanger is preferred.

6.4 USE OF SODIUM TOLUENE SULFONATE

Sodium toluene sulfonate is used in synthetic granules to improve several processing and finished product characteristics. The benefits of using STS are a result of its ability to:

(a) Reduce viscosity and surface tension of acid mix, paste, and crutcher mix.
(b) Improve puffability.
(c) Improve product crispness and resistance to caking.

The reduced viscosity and surface tension result in better atomizability and, therefore, decreased particle size. The smaller particles, however, can be highly puffed so that density may remain low. There is some evidence to show that this increased puffability results from the ability of the mix to hold more air. More often than not, however, STS is added to increase product density rather than to decrease it. The reduced viscosity allows lower crutcher mix moistures for the same atomizability. Steam puffing is thus reduced and a density increase results.

One of the major consequences of using STS is an increase in tower rates. Increased tower rates are due to:

(a) Less agglomeration since granules are crisper.
(b) Reduced tower evaporative load since lower crutcher mix moistures are possible and higher finished product moistures are obtainable with the crisper product.
(c) Increased thermal efficiency of the tower. If the crutcher mix atomizability is improved (a balance between STS and reduced crutcher moisture), the thermal efficiency of the tower will be increased due to the larger surface area for heat and mass transfer obtained with the smaller particles. This is reflected in lower exhaust air temperature.

6.5 AIR PHASE IN CRUTCHER MIX

Air in a crutcher mix is an additional phase. Air can enter crutcher mix in several ways. It may be present in the paste, it can be pure STPP hexahydrate, which is a free flowing, apparently dry solid. Thus, while free moisture in the finished granules can lead to softness or stickiness, any water bound as hexahydrate can be tolerated and, since it can replace sodium sulfate in the formula, represents a cost advantage. In addition, more water left in the product means less to be removed during spray drying, carried in with the addition of dry solids, or it can be physically beaten into the mix whenever the crutcher agitator is on.

Because of the large effect of air puffing on the density of the dried particles, the air content of crutcher mix must be closely controlled. For most formulations, the finished product density requirements are such that a special effort to add or remove air is not necessary. With these formulations, the most important thing is uniform operation in paste making and crutching. Excessive agitation of paste or crutcher mix, wide variations in scrap usage, etc., can cause wide variations in crutcher mix air content and consequently poor density control in the tower. With perborate containing brands, excessive scrap usage can cause severe crutcher mix aeration as the perborate decomposes in the crutcher.

A rapid and simple check on crutcher mix air content is to measure the density of the crutcher mix. Normal crutcher mix will have a density of 1.0 to 1.05 g/cc. For some formulas, air must be either added or removed from the mix in order to meet the finished product density requirements.

6.5.1 Air Additions

To lower particle density, air may be added to a crutcher mix by increasing crutcher agitation, or by using higher levels of scrap. These methods can add only a limited amount of air and cannot be controlled closely.

For more precise control, air may be injected into the mix as it is pumped to the nozzles. Air injection can lower product density by 0.03 g/cc. Strictly speaking, air injection is related more to spray drying than to crutching. The amount of air that can be effectively added to a crutcher mix is limited. Excessive air may result in either high coarse scrap rates or in a density increase due to a shattering of the particles. Dusty product may also result.

6.5.2 Deaeration

An increase in product density can be achieved by removing the air in a crutcher mix. Deaeration can be done either mechanically or chemically.

6.5.2.1 Chemical Deaeration

Materials such as sodium chloride and some non-ionics will change the normal lye-in-neat emulsion of crutcher mix to a continuous lye phase with islands of neat. The greatly reduced viscosity of this phase structure allows the air which concentrates in the lye phase to escape. To be effective, the chemical must be added before the phosphate is added to the mix, and a quiescent period (no agitation) must follow the addition. Crutcher mix densities of up to 1.4 g/cc are obtained by chemical deaeration.

Salt is commonly used in high active, unbuilt brands which tend to spray dry at low densities. The high sulfate content of these formulas allows the substitution of salt at the required 10% level. If nonionics are used, their surface-active properties must be considered since they will also affect both sudsing and detergency.

6.5.2.2 Mechanical Deaeration

By flashing crutcher mix through a nozzle at high temperature, or under a vacuum, air is stripped from the mix. Densities equivalent to those from chemical deaeration are obtained. Although no reformulation of the product is required, as is necessary with chemical deaeration, extra equipment is needed.

Examples:
Some of the fundamental principles thus far discussed may become clearer by considering two special cases.

1. Increased atomizability at constant puffability
   Under these conditions, both atomizability and puffability are in the direction of increased density. In general, increasing atomizability at constant puffability will increase product density and reduce particle size. This can be done with a low free moisture crutcher mix by heating the mix below 100 °C. As the temperatures rises above 100 °C, however, density will decrease because the free moisture will flash to vapor at the nozzle and puff the granules. With still higher temperatures, the product density will increase again due to the shattering of the particle; however, this normally results in excessive dustiness.
   With high free-moisture crutcher mixes, atomizability is already good and, therefore, the effect of increasing mix temperature is small. In fact, if steam injection is used, the additional free moisture of the condensed steam may lower density by increasing steam puffing.
2. Changing puffability constant atomizability
   If puffability is increased, product density is reduced and particle size is increased. Air injection or mechanical deaeration are excellent examples of changing puffability without affecting atomizability. Chemical deaeration, however, affects atomizability as well as puffability since crutcher mix viscosity is also reduced.

The effects of phosphate hydration and STS level cannot be predicted as easily. Atomizability and puffability increase or decrease together when these variables are changed.

Depending on the formula, equipment, and processing conditions used, a change in phosphate hydration or STS level may increase, reduce or make no change in density and particle size of the finished product.

6.5.2.2.1 Processing

The crutching system consists of the tanks and bins required to feed the various raw materials, the “crutcher” in which these materials are mixed and the “drop tank” into which the crutcher mix is discharged from the crutcher and from which the mix is fed to the spray-drying operation.

6.5.2.2.2 Batch Crutching

A standard batch crutcher is designed whereby mixing is accomplished by a screw-and-sleeve type agitator. Cylindrical sleeve with a slot cut-out near the top is mounted at the center of the crutcher, stopping short of the bottom. A screw-type agitator turns within this sleeve. The end of the screw widens out so as to sweep the bottom of the crutcher. Crutcher mix is pumped through the sleeve by rotation of the screw, thus mixing materials as they are added. By reversing the rotation of the screw, the direction of flow in the crutcher may be reversed. The batch crutcher is usually scale mounted, so that all major ingredients may be weighed as added. If the crutcher is not scale mounted, all ingredients must be pre-weighed or premeasured before adding to the crutcher.

Crutcher mix is pumped through the sleeve by rotation of the screw, thus mixing materials as they are added. By reversing the rotation of the screw, the direction of flow in the crutcher may be reversed.

6.5.2.2.3 Materials Mixing

Paste is always added to the crutcher first. Other ingredients are added to the crutcher in an order designed to give optimum mixing and fluidity consistent with the required level of hydration. The mixing action of the crutcher during materials addition is as follows:

Before the paste level is up to the sleeve, the movement of the sweep in the bottom of the crutcher gives good mixing. However, as soon as the level reaches the bottom of the sleeve, the surface of the paste is quiescent. The screw is unable to lift the fluid paste to the top of the sleeve so that if minor ingredients are added at this time, they will tend to lump and not be mixed into the paste.

As soon as the level is up to the bottom of the slot, material can be drawn into the screw running down (reverse) and mixed into the batch. Dry material added at this time is incorporated satisfactorily if added just outside the slot. The screw is run down during STPP addition because the STPP would otherwise be forced to the sides of the crutcher where it would cake.

After addition of dry solids, mixing is again fairly good. The paste is then stiff enough to be lifted by the screw so that it flows over the top of the sleeve. Thus, the general procedure is to pump the mix up through the sleeve when wet materials are being added, to achieve a splashing action on the top of the crutcher. When dry materials are added, the material is drawn down through the sleeve to wet and disperse them promptly before lumps can be formed. When all the ingredients have been added, the level in the crutcher should be about 10 cm above the level of the draft tube, so that there will be a good flow of mix into the agitator from all sides. This provides the optimum conditions for good blending during final mixing.

Most mixing problems arise from capacity requirements. A common problem is an undersized agitator motor. A motor which may have been quite satisfactory at low rates will not handle the drier, more hydrated crutcher mixes are required to increase rate. As production rates are increased, crutching time often becomes a bottleneck. When this happens, there is a tendency to try to get by with less mixing and product quality often suffers. The way to increase crutcher capacity is to reduce material addition time. Increasing paste transfer rate or pre-weighing ingredients such as silicate can save valuable minutes.

Poor mixing can also be caused by improper batch sizes. The level of liquid ingredients in the tank must be high enough to ensure good circulation through the slot before any dry ingredients are added. The crutcher must not be overly full at the end of addition, however, or agitation will be repressed.

6.5.2.2.4 Weighing

To control product composition, the required quantity of each ingredient must be carefully weighed into the crutcher. Deviations generally occur when either a double addition or no addition at all of a particular constituent is made. For this reason, a record of the cumulative weights, as well as a record of the individual weights of ingredients added, is kept. When deviations in product composition occur, the crutcher card should be checked to be sure that it corresponds to the formula card in use. Where crutcher cards are prepared for two or three different formulations or batch sizes, these should be kept on separate cards and only one card at a time issued to the operator.

The crutcher scale itself provides sufficient accuracy for the addition of major ingredients. Some materials, such as amide, which are added in intermediate quantities are best weighed by stopping the agitator near the end of the addition, or by premeasurement by weight or in a gage tank. The minor ingredients (CMC, brightener, etc.) are all pre-weighed on a sensitive balance. One potential source of error other than a weighing error is failure to close valves completely so that materials may slowly leak in during subsequent weighing.

6.5.2.2.5 Solids Buildup

Solids will build up on the upper parts of the crutcher side walls when dry material is thrown outward by the agitator. Splashing liquid and pieces of dry scrap which wet out slowly will also stick to the sidewalls.

Buildup in the dry materials chutes can cause the material to fall on the agitator itself rather than on the liquid surface. Material hitting the agitator will be thrown to the sidewalls. Regular inspection and cleaning of these chutes will eliminate this cause of buildup.

Regular scraping of the sidewalls will maintain the effective size of the crutcher and prevent changes of the batch level in the crutcher. Scraping should be frequent to minimize the amount of aged product going into any one crutcher. Scraping should be done early in the cycle to insure dissolution of the lumps.

6.5.2.2.6 Moisture

Variations in the moisture content of the finished crutcher mix is affected by variations in paste moisture and by the quantity and moisture level of wet and dry scraps. These variations are minimized by controlling the paste moisture content and by attempting to use the same combination of scraps for each batch.

6.5.2.2.7 Aeration

Crutcher mix is subject to air entrainment from many sources such as leaky pump packings, aeration during crutching, and air addition with dry materials or with paste. Badly aerated crutcher mixes cause difficulty in pumping, in spraying, and in finished product density control.

When the crutcher agitator is operating downward, as it must be when dry ingredients are added, a great deal of air is whipped into the mix. The amount of downward operation of the screw should, therefore, be carefully controlled.

As with moisture control, the same combination of scraps should be used in each batch. It is undesirable, for example, to make up one crutcher batch with all fines only and the following one with all coarse scrap.

6.5.2.2.8 Material Addition

Paste is pumped into the crutcher from the paste surge tank. Excessively long paste transfer time can result from high paste viscosities due to low moisture, insufficient toluene, or excessive aeration in the surge tank.

STPP is added to the crutcher from a bin above the crutcher, either by gravity or by a screw conveyor. Excellent mixing of STPP is essential if a lump-free crutcher mix is to be obtained. While the addition of STPP may be very fast or slow, depending on equipment and cycle time, it must in all cases be uniform. Uneven slugs of STPP are a major cause of lumping and crutcher buildup.

Silicate is added from an overhead work tank. Silicate is normally added to the crutcher at ambient temperature; for special formulations, it may be heated by a coil in the work tank.

Sodium Sulfate. In some formulas, much more sodium sulfate is required as a filler than is available in the paste. Solid sulfate is, therefore, added to the crutcher, either from a bin or from bags. Addition of these large amounts of sulfate can cause pumping problems, severe strainer and nozzle makeup, and excessively high coarse scrap rates. Good mixing of the sulfate into the mix has been found to be most important. The sulfate should be added to the crutcher immediately after the paste, to maximize mixing time. One plant with a crutcher-neutralizer has obtained a large improvement in processing by adding the sulfate to the caustic and water heel before neutralization. Another plant has eliminated severe lumping and makeup problems in its high sulfate formulas by using a coarser sulfate. This shows that physical form is also an important factor.

6.5.2.2.9 Brighteners, Pigments, and Dyes

In the final crutcher mix, the distribution of major ingredients throughout the batch is accomplished reasonably well but agitation is not intense enough for good dispersion of certain minors. Lumps or agglomerates of brighteners, dyes, and pigments in the finished product cause staining or spotting of fabrics during use. These materials, therefore, are usually pre-dispersed prior to being added to the crutcher. Pre-dispersion effectively breaks up lumps and, in addition, aids in the more uniform distribution of the materials throughout the mix.

Pre-dispersion is accomplished in a small tank set above the crutcher. It is equipped with a high shear agitator, such as a dispensator, which is designed to provide both good mixing of dry materials with liquid in the tank and also to break up any agglomerates or lumps of the solid material. Paste has been found to be a universally good pre-dispersion medium. Water and silicate have also been used, but they may cause lumping when used with certain brighteners.

Dyes are water-soluble and are normally dissolved before being added to the pre-dispersion tank. The concentration of dye solution used is not critical provided only that complete solution of the dye is obtained. This solution is filtered as insurance against an occasional undissolved particle.

Pigments are not water-soluble and are believed to be the major cause of fabric staining. Pigments are added directly to the pre-dispersion tank and from there to the crutcher. Pigments are currently used in either powder or paste form. The paste form has the advantage of being easier to handle in the plant.

6.5.2.2.10 Other Minors

Other minor ingredients, such as CMC, EDTA, and BTA, are pre-weighed on a sensitive balance and then added to the crutcher. With these minors, normal crutcher agitation will give satisfactory dispersion if they are added early in the crutching cycle. The pre-dispersion tank can be used if necessary.

6.5.2.2.11 Amides, Fatty Acids, and Nonionics

Amides, fatty acids, and some nonionics can be handled in either solid or liquid form. To be handled as a solid, the material must be available in a suitable physical form such as flakes, chips, or powder that will dissolve or melt and disperse uniformly throughout the crutcher mix in the limited mixing time available. To be handled as a liquid, the material must be melted and stored without significant degradation or discoloring.

The advantage of using the material in a solid form is that no additional handling equipment is required. The material can be simply dumped from bags into the crutcher. In general, a material will be handled in its solid form if it is not frequently used.

Liquids, on the other hand, require less storage space than bulk solids, are easier to transfer from storage to the crutcher, and can be added more rapidly. In general, a material will be handled as a liquid if it is frequently used, if it can be alternated with other materials in the same handling system, or if the advantage of quick addition will justify the extra equipment expense.

Flaked amide can be added to the crutcher at normal crutching temperature. Addition should be early in the cycle to provide time for compete dispersion. If liquid amides are added to a crutcher of too low a temperature, however, the amide will freeze into lumps in the crutcher mix. Dispersion and, therefore, product performance will be poor. Amides tend to darken on aging, particularly at elevated temperatures. Therefore, care must be taken to avoid overheating amide which is in direct contact with steam coils. Ammonia amides in storage must be kept dry since they hydrolyze readily.

6.6 AUTOMATIC BATCH CRUTCHING

Operation of the batch crutcher can be made completely automatic with sufficient instrumentation. Materials addition, agitator operation, and crutcher mix withdrawal are completely controlled by programed instruments. Automatic batch crutching provides more uniform operation from batch to batch and reduces labor cost in plants where the crutcher operator can be eliminated.

6.7 CONTINUOUS CRUTCHING

Because of its cyclic nature and of its slow response to change, the batch crutcher causes variations in the mix fed to the tower. Continuous crutchers have been developed, therefore, which produce crutcher mix continuously and can be controlled easily and rapidly to provide greatly improved mixing.

All of the crutcher mix ingredients, both wet and dry, are continuously metered into a mixing and wetting tank having a high level of agitation. From here, the mix obtained overflows into an aging vessel which controls hydration time. A level controller senses the level of the product in the aging tank and regulates the rate of addition of each of the ingredients accordingly. The cost of such a unit is high, however, and the quality improvement and possible labor savings do not normally justify its installation.

6.8 DROP TANK

The crutcher is emptied by dropping the charge through a strainer into the drop tank. During the dropping operation, a small quantity of steam is often bled into the top of the crutcher to improve draining of the crutcher mix from the walls of the crutcher. The strainer removes tramp metal and other objects which might interfere with subsequent operations. The drop tank serves two functions:

(a) It provides surge capacity between the batch crutching and continuous spray-drying operations.
(b) It allows crude control of crutcher mix aging time.

The drop tank usually holds approximately two crutcher batches. It may be provided with a screw agitator, similar to the one in the crutcher, capable of supplying a low level of mixing.

At some of the small plants, the drop tank is at the same or at a higher level than the crutcher and, therefore, a pump is required to move the mix from crutcher to drop tank. Often this crutcher mix transfer is the rate-limiting operation.

6.9 INSTRUMENTS TO MONITOR CRUTCHING

Both crutcher mix temperature and crutcher agitator loading vary with time from beginning to end of the crutching cycle. By monitoring these variables, valuable information, especially the batch-to-batch variability, can be ascertained. Recording instruments with eight-hour charts have been found to work best. The scale width should be large enough to provide an accurate analysis of the trace. The instruments used are:

(A) Temperature Recorder

This follows the crutching temperature during the cycle. Initially the temperature recorded is the paste temperature. As each ingredient is added, the temperature rises or falls depending on whether the material is colder or hotter than that in the crutcher and depending on the progress of phosphate hydration.

(B) Wattmeter Recorder

As each material is added to the crutcher, the power drawn by the crutcher agitator increases. Phosphate hydration increases the power consumption even further since the viscosity of the mix increases as free moisture is withdrawn.

Watt meters are preferred to ammeters since they are not sensitive to line voltage and power factor variations. Instruments can also be installed on the drop tank to follow "after crutching" temperature and agitator loading.

7 Spray Nozzles

Spray nozzles are used both for atomizing crutcher mix in the spray-drying operation and for spraying onto the dried granules materials, such as perfume, magnesium sulfate solution, etc., which cannot be added satisfactorily in the crutcher. They are also used in specialized applications, such as agglomeration and coating of various granular materials. Several types of nozzles are available and the selection of the proper spray system for a given service will depend upon:

1. The physical properties of the material to be sprayed
2. The degree of atomization necessary
3. The rate of spray desired
4. The equipment available that will handle the desired capacity

The physical properties of materials such as viscosity, surface tension, freezing point, and volatility obviously must be considered in designing a spray system, as well as the degree of atomization that will be satisfactory. Too great a viscosity will generally produce a coarse spray. However, with the materials that are used, viscosity can usually be controlled to the desired level (e.g., by increasing temperature and/or moisture of a crutcher mix) to give a good spray. Freezing can be a problem with some relatively high melting point materials (e.g., fatty acids, amides) if the proper application temperatures are not maintained. Over atomization, particularly of volatile materials, will lead to losses and a problem of fume disposal.

The two basic types of nozzles that are normally used are two-fluid and pressure nozzles. The degree of atomization in either application is dependent, within certain limits, upon the energy put into the fluid. In the case of two-fluid atomization, the energy necessary to form the spray is largely supplied by the expansion of a compressed gas, usually air, which entrains the liquid to be sprayed as the gas attains velocity by expanding through an orifice. The energy necessary for atomization with the pressure nozzle is supplied solely by the fluid to be atomized. The fluid under pressure is discharged through an orifice and attains sufficient velocity to be atomized by the conversion of potential energy to kinetic energy as the pressure drops across the orifice.

DOI: 10.1201/9781003596219-8

Pressure nozzles are used for spray-drying granules and for spraying high melting point liquids which would freeze if exposed to cold air in a two-fluid nozzle. Two-fluid nozzles are usually used for applying liquids to the surface of granules; MgSO, perfume, and dye spray-ons are typical applications.

7.1 TWO-FLUID ATOMIZATION

The two-fluid or air atomizing nozzle as we use it gives better atomization than pressure nozzles at any given rate; however, its field of practical application is limited to relatively low flow rates (0.05–5 kg/min, vs. 10–50 kg/min, for the pressure nozzle). The liquid is usually metered to the nozzle at a pressure of less than 2 kg/cm and is dispersed by the high velocity stream of gas. This system will normally give the smallest drop sizes in spraying and is employed, for example, for spray-ons when low flow rates are encountered or when it is necessary to achieve a high degree of atomization in order to secure the most complete coverage of the granules. There is the advantage of not requiring a high-pressure pump to achieve atomization but the disadvantage of requiring a source of compressed air (up to 1 lb. of air per lb. of liquid to be atomized). This system also has the advantage of flexibility in allowing rate changes without greatly changing the degree of atomization.
There are disadvantages in the use of the two-fluid nozzle:

1. There is a tendency toward misting and loss of material if the pressures are not well controlled. Misting will also lead to a fume disposal problem.
2. An exhaust system must be provided to handle the volume of atomizing air if the spraying is done in a closed system.
3. Usually, the spraying must be done on cascading granules, as in a rotating drum or a perfume box, because the air has a tendency to blow granules away from the spray area when the spray is applied to a moving belt and also because efficiency of coverage is much better when the spray is directed at cascading granules.

When spraying high-melting or sticky materials, there is a tendency for product buildup on the nozzle and partial or total makeup of the orifice with consequent deterioration of the spray. Periodic inspection and cleaning is required in such cases.

7.2 PRESSURE ATOMIZATION

The use of pressure nozzles will give a good degree of atomization over a wide range of flows with less chance of misting as long as pressure is properly controlled. The energy input for atomization is provided only by the high velocity of the liquid stream as it is forced through an orifice into the surrounding atmosphere. For a liquid of given physical characteristics, the pressure upstream of the orifice and the orifice size determine the discharge rate and the degree of atomization. As the pressure upstream is increased, the velocity of the stream is increased and atomization is attained as more potential energy is converted into kinetic energy. The amount of particle breakup increases as pressure and jet velocity in turn are increased and too great a pressure will lead to misting which is generally undesirable. If the pressure is too low for a given system, there will not be sufficient energy supplied to form a spray and a single stream or partial spray may be formed.

It is possible to predict the atomization pressure for a given system if enough of the variables for the fluid are known. For example, the Reynolds number, Re, is plotted versus a dimensionless group, Z, whose importance can be predicted by dimensional analysis. Z is correlated with:

d = nozzle orifice diameter
v = velocity of the fluid
ρ = density of the fluid
μ = absolute viscosity
σ = surface tension

The plot is divided into four areas; fourth area of gravity flow (dripping). Area I in the plot is the region where decomposition of the single stream takes place due to surface turbulence. As pressure, hence velocity, increases, Area II is reached where the decomposition of the stream increases with more symmetrical helical turbulence. In Area III, the pressure has increased to the point where the velocity is sufficient for atomization to take place. From the plot, it can be seen that with the increase in Reynolds number due to increase in velocity, the various stages lead to atomization.

Pressure atomization will generally give a somewhat lesser degree of atomization than will two-fluid atomization. In the case of spray drying, pressure nozzles provide just about the desired particle size, whereas two-fluid nozzles produce too fine a spray and thus excessive dust.

Pressure nozzles are less prone to over atomization and loss due to misting. When used for spraying onto solids, there is good penetration of the spray and less displacement of granules compared with air atomization. Pressure nozzles are often used to spray high melting liquids which cannot be air-atomized due to the cooling effect of the air.

The pressure nozzle will not normally provide as complete a coverage of the granules as will air atomization and is usually not as suitable when highly efficient coating is desired. Since atomization and rate both change with pressure, it is more difficult to maintain the same degree of atomization as rate is changed. It is usually advantageous to use a number of pressure nozzles of low capacity to give sufficient flow at the desired pressure range for good atomization. These nozzles are available in wide capacity and pressure ranges. Changes in rate can be achieved with the least change in pressure and atomization by adding or subtracting nozzles.

The maintenance and inspection of pressure nozzles is important since a fair degree of erosion occurs. If the orifice or internal chamber becomes worn or nicked, atomization will suffer. Normally, the nozzle tip and whirl chamber should have the same diameter. If a narrower cone angle is desired, the whirl chamber can be one or two sizes (6th of an inch) larger. A large tip with a small chamber, however, will give too wide a spray cone and cause rapid erosion of the nozzle.

There are two basic types of pressure nozzles: the hollow cone, and the solid cone.

7.2.1 Hollow Cone Pressure Nozzle

The hollow cone spray pattern is formed by introducing the fluid into the chamber in a direction perpendicular to the direction of spray and causing a centrifugal action to be set up within the chamber. This centrifugal motion continues as the fluid is discharged through an orifice and results in the hollow cone pattern as the droplets are forced outward by the centrifugal force.

This nozzle provides the finest atomization in relation to pressure and capacity of any of the pressure nozzles. There is less tendency to clog with this type as it has the largest orifice diameter for a given capacity and no internal obstructions. Although this nozzle does not run full, the discharge rate is proportional to the orifice area and the pressure. Spray drying is most often done with this type of nozzle.

7.2.2 Solid Cone

The solid cone pressure nozzle provides a spray pattern over a circular area. This pattern is achieved by allowing the fluid to enter the nozzle in the same direction as the spray discharge, pass through an internal vane or baffle and a long chamber which produces turbulence without enough centrifugal force to form a hollow cone.

The solid cone nozzles at a given capacity have a smaller discharge orifice than do hollow cones. This smaller orifice and the internal vanes give these nozzles a greater tendency to clog than hollow cone nozzles. The spray angle is also narrower for solid cone nozzles as compared to hollow cones at the same pressure. These nozzles are not normally used for spray drying except in the pilot plant towers where the smallest hollow cone nozzle available is still too large for the low rate.

7.3 SCALE-UP OF PRESSURE NOZZLES

With a pressure nozzle, the spraying rate is increased by increasing spraying pressure. As has already been mentioned, however, a change in pressure will also change the degree of atomization. In addition, spraying pressures are often determined by equipment limitations. In practice, therefore, in order to increase rates, either additional nozzles or nozzles of a larger size are generally used. The maximum number of nozzles which can be used simultaneously is usually limited by the requirement that impingement of the spray cones on each other and on the tower, drum, or spray-on box walls be minimized. That is, the number of nozzles is limited, for example, by the tower diameter or by the drum length and diameter. On the other hand, changing to a larger size nozzle at the same operating pressure generally increases the average particle size of the spray produced.

Thus, when increasing rates, both number of nozzles and nozzle sizes are normally increased to give the best compromise in product characteristics. However, it has been found, that two nozzles which are geometrical scale-ups of each other, that is, which are scaled-up in such a way that the ratios of all the nozzle dimensions remain constant, will produce the same degree of atomization at any given spraying pressure. This means that it is possible to scale up a standard pressure nozzle in such a way that, at any given pressure, the scaled-up nozzle will spray at a higher rate than the standard nozzle and yet produce droplets of the same particle size distribution. The nozzle selected for scale-up is usually one which is known to produce droplets of the desired size at the available pressure. The scale up is based on the process variables influencing the spraying rate Q and the orifice diameter. Thus, for example, in order to double nozzle capacity, the scale-up factor affects every dimension in the scaled-up nozzle, which must be multiple times the corresponding dimension in the standard nozzle.

7.4 SCALE-UP OF NOZZLES

Work to date indicates that in spray-drying towers, drying capacity is reduced by approximately 10% when a single, large capacity nozzle is used in place of an array of conventional nozzles. This is believed to be due to a combination of two effects:

1. A single hollow spray cone does not distribute granules in the center of the tower.
2. The effective tower height is reduced by the deeper penetration of a single spray cone.

7.5 SPINNING DISK ATOMIZATION

Atomization by the spinning disk technique differs from two-fluid or pressure atomization in that it does not require nozzles or orifices. The liquid is introduced at the center of a rapidly spinning disk and breaks up into droplets as it leaves the periphery of the disk. Some control of particle size can be achieved by varying the speed of rotation. The advantages of spinning disk atomization are:

1. No high-pressure pump is required. Liquid feed to the disk can be at essentially atmospheric pressure.
2. No problems of nozzle plugging or nozzle wear.
3. Greater flexibility in varying feed rate.
4. Greater uniformity of droplet size.

The drawbacks stem mainly from its mechanical features, i.e., the difficulty of achieving high rates on one disk and the difficulty of locating more than one disk in a tower. Since there is a possibility of wall build-up because of the radial path of the spray, they are generally used in conjunction with large diameter, concurrent, spray-drying towers.

8 Spray Drying

Spray drying is essentially the final step in synthetic detergent processing. The crutcher mix is atomized in the spray tower and the droplets formed are dried as they fall through a turbulent stream of hot air. The dry granules are collected at the base of the tower and, although they undergo one or more subsequent spray-on and size classification operations, at this point they have assumed their final physical characteristics.

Since product appearance is a very significant factor in customer appeal, and product density directly affects the packing operation, spray drying becomes more than merely a drying operation. Product moisture is only one of the many quality items controlled in the spray-drying process.

8.1 SPRAY-DRYING MECHANISM

Exact knowledge of what occurs in the spray-drying tower is a process of continuous improvement. Present approach is limited to a probable mechanism based on measurements of spray-drying variables. Four zones in the spray-drying process are postulated.

8.1.1 Particle Forming Zone

The atomizing nozzles spread the crutcher mix into a thin conical sheet. This sheet becomes thinner as it spreads away from the nozzle and finally disintegrates into droplets. The initial size, shape, temperature, and moisture of the droplets are determined both by the spray nozzle and by the nature of the mix.

As the mix leaves the nozzle, free vapor is released and air puffing takes place. If the crutcher mix contains excessive air or if the crutcher mix temperature is significantly above the boiling point, fragmentation of the droplets may also occur.

8.1.2 Surface-Drying Zone

During the surface-drying stage, moisture evaporates from the surface of the granule with no change in granule diameter. As long as the rate of diffusion of moisture from

DOI: 10.1201/9781003596219-9

the interior of the granule to the surface is sufficient to provide enough water for surface evaporation to balance the rate of heat transfer, granule temperature will remain constant at the tower wet bulb temperature.

The rate of diffusion depends on mass transfer within the granule and drops rapidly as the granule loses moisture. The rate of heat transfer, on the other hand, remains relatively constant throughout the residence of the granule in the tower. When the rate of surface evaporation no longer balances the rate of heat transfer, the temperature of the particles will increase.

8.1.3 Puffing Zone

As the temperature of the granule increases, the vapor pressure of water within the granule increases and steam puffing begins. Puffing continues until the surface has lost sufficient moisture to form a solid crust and will not expand further.

Finished product density is controlled by the puffing zone. At this stage, a dynamic balance exists between the rates of drying and puffing. Any change whose dominant effect is to increase heat transfer decreases density by increasing puffing. A change that primarily increases mass transfer increases density. For example, an increase in air temperature increases heat transfer and decreases density. An increase in swirl (turbulence) increases mass transfer more than heat transfer and increases density. Changes which shorten the length of the puffing zone increase density and vice versa. Thus, an increase in hydration, for example, with no other change, decreases free moisture, giving a shorter puffing period with a resultant increase in density.

8.1.4 Finishing (or Falling Rate Zone)

Here, heat and mass transfer within the granule are both low, and surface drying, limited by diffusion, continues at a decreasing rate. If heat transfer exceeds mass transfer during this period, steam pressure will create blowholes in the surface, but no puffing occurs.

The falling rate drying zone determines final product moisture and, probably, the extent of phosphate reversion. The length of this zone is determined by the difference between the time required to form a solid crust on the granules and the tower residence time. Thus, factors such as increased tower temperature or decreased product feed rate normally extend this zone.

Not all particles pass through all these steps, primarily because of differences in particle size. Very small particles lose moisture so rapidly that they are too dry to puff before they are hot enough. Extremely large particles fall rapidly and heat slowly and thus never reach puffing temperature.

8.2 DRYING VARIABLES

The instantaneous rate of drying of any single particle is a function of two classes of variables: (1) variables in the particle itself, and (2) variables in the environment. For a given environment, the rate of drying of a particle is dependent upon the moisture content and distribution, the size and shape, the temperature, and the nature of the material.

The environmental factors influencing rate of drying are the temperature and the relative velocity of air and particle. Other factors are theoretically important but have no practical bearing on making granules.

Both temperature and air particle velocity are tied in with the air flow pattern in a tower. The most important factor is turbulence. This is not a gross turnover but is a localized, random, transitory motion induced by fluid shear. Turbulence is not synonymous with swirl, but control of swirl offers a means for regulating turbulence. Turbulence results in mixing, and thus increased turbulence decreases the temperature gradients in a tower. This effect is so great that the average air temperature in a tower is much closer to the exhaust temperature than it is to the inlet. It may be seen that the actual temperature in the tower is considerably lower than the inlet air temperature. This is due to mixing. The exhaust temperature is lower than the inlet by the amount of heat used for evaporation (and some heat loss) and is thus dependent upon the drying variables. Although tower exhaust air temperature is a dependent variable, it is a key factor in determining the effects on drying rate and density of changes in the independent variables.

Moisture, density, and rate are controlling considerations in spray drying of synthetic granules. Thus, it is convenient to consider the spray-drying process in terms of these three variables. The relationship among the three may be expressed graphically by a "tower operating curve". Existence of such a curve may be demonstrated by the following reasoning:

(a) Consider a tower operating at fixed rate, fixed feed conditions, and fixed inlet air conditions including temperature and air throughput. Such a tower would have a product with a certain density and moisture, and neither the density nor the moisture would change as long as all conditions were held constant. This situation may be represented by a point on a graph of moisture vs. density.

(b) If the rate of feed of crutcher mix to the tower were increased with no change in other independent variables, there would be more droplet surface in the tower. With more surface for evaporation, there would be more total drying, and therefore, the tower air temperature would be lower. (Tower air temperature is a dependent variable and could not be held fixed.) Because of the lower temperature, there would be less drying from each particle, and therefore, product moisture would increase. Also, since the drying would take place at a lower temperature, there would be less steam puffing and, hence, a higher product density. This represents a second point on the plot of moisture vs. density.

(c) Continuing this analysis for a series of rates, there would be a corresponding series of values of moisture and density, other conditions still fixed. Drawing a curve through these points would give the density-moisture relationship for this particular set of conditions. Each point on this curve would correspond to a certain rate. By adding a third axis, rate, a three-dimensional curve would be obtained.

This curve, as described, is specific for one set of conditions. However, the effects of other variables may be deduced by considering the effects on moisture and density of changing each of the other variables at constant rate. Since the effects would be directionally the same at any reasonable rate, the net effect of the change would be to shift the operating curve. The operating curve has been made available in the form of the "spraying drying model". This model represents density and normal rate. For example, a higher inlet air temperature will result in a lower moisture and a lower density, but the relative magnitudes of the two effects are approximations. The examples that follow will show the reasoning involved in deducing these probable effects.

The effect of drop size may be considered by assuming a coarser spray at constant rate. This would give less surface in the tower for evaporation. Thus, there would be less drying and a higher air temperature. Less drying would result in higher product moisture. The higher air temperature would cause more steam puffing and result in lower product density. This would shift the operating line in the direction of a lighter, wetter product.

As a second example, a higher crutcher moisture would from heat balance result in a higher product moisture, but there might also be effects on density. With a higher particle moisture, there would be more evaporation, and therefore, a lower air temperature, tending to raise density. However, the higher moisture would tend to increase puffing and lower density. The net effect would be small and might be in either direction, so only the principal effect is emphasized. (This, of course, assumes that the change in crutcher moisture was small enough so that atomizability, which affects drop size and agglomeration, is not appreciably affected. When major changes of a variable are considered, these side effects can often greatly influence the overall effect.) The operation varying rate only is along the tower operating curve. For the product to be acceptable, it must also be at proper density.

Changing a variable would shift the operating line and will, in the general case, result in a different density and moisture at the same rate. For example, suppose the powder began to blow light but at the right moisture. This indicates a shift in the operating curve. The density could be corrected by increasing rate but the moisture would also go up. An increase in inlet air temperature would probably correct the moisture back to the value of the variable increase target and have little effect on density since the exhaust air temperature would be practically unchanged.

Although the spray-drying model helps in the understanding of spray tower operation, it is by no means the whole story. Thus far we have made little or no distinction between dependent and independent variables or between controllable and non-controllable variables. In a subsequent section, we will see what controls can be used and how they affect the principal finished product characteristics.

- Moisture
- Puffability
- Air injection
- Air temperature
- Throughput

- Drop size atomizability
- Crutcher moisture
- Turbulence
- Height
- Agglomeration

8.3 PROCESSING EQUIPMENT

Most plants, regardless of whether they have batch or continuous sulfonation units, have similar spray-drying systems which differ primarily in size of the tower. The exception, a co-current tower, is discussed later. A flowchart of a typical spray-drying system usually presents a clearer picture of the operation. It is convenient to consider the product and hot air flows separately.

8.3.1 Product Flow

Crutcher mix normally contains moisture and has a viscosity of about 1,000–50,000 p. Its temperature is generally between 65 °C and 90 °C except in a few cases where it is heated further in order to obtain special, high-density product. This extra heating is accomplished either by a special heat exchanger or by direct steam injection.

Crutcher mix is transported from the drop tank or crutcher mix storage tank by a low-pressure pump which forces the slurry through a strainer and provides a positive feed for the high-pressure pump. If a steam injection system is installed, a rotary pump is used to feed the low-pressure pump. The high-pressure pump delivers the crutcher mix to the nozzles at the top of the tower, where atomization takes place at pressures of 258 kg/cm^2. The droplets are dried as they fall through the tower and the finished granules are removed from the bottom of the tower cone. In some plants, the granules leave the tower through a feeder valve which serves as an air lock. At the base of the tower, the granules enter the product handling system.

8.3.1.1 Low-Pressure Pump

The low-pressure pump is usually a large, slow, reciprocating steam pump whose capacity is varied by changing the steam flow and thus decreasing or increasing the pump speed. Pressures generated by the pump are usually between 7 and 20 kg/cm^2. Much of this pressure is used in the straining operation, leaving 3 to 9 kg/cm^2 at the suction of the high-pressure pump. This positive pressure is necessary to ensure that the cylinders of the high-pressure pump are filled on the suction stroke. The discharge pressure of the low-pressure pump varies appreciably due to the reciprocating action. In some units, these fluctuations are reduced by the use of a pulsation dampener consisting of a pressure chamber with an air bladder. In any case, it is considered advisable to use "snubbers" and diaphragms on the pressure gages to dampen needle fluctuations and prolong gage life.

Flowchart of Counter-Current Spray-Drying System – See Appendix

Exhaust fan
Scrap
Nozzles
Stack
Tank
Strainers
Plenum
Chamber
Tower
Rotary pump
Steam injection
Smaller towers only
Granules to product handling system
Forced air
Fan
Air

Low-pressure pumps are generally quite dependable with the chief problem being occasional instances of foreign material in the valves. Most of this material should be removed by the strainer between the crutcher and the drop tank. Magnetic "nail traps" are sometimes used for further protection. Safety from overpressure is provided by a rupture disc or relief valve.

In plants where steam is not available, an electrically driven low-pressure pump, either rotary or piston type, is used. Unlike steam-driven pumps that are self-regulating, the electrically driven pump runs at constant speed. Because of this, some crutcher mix must be continually bypassed through either a manual valve or the pressure relief valve to prevent extreme pressure fluctuations.

8.3.1.2 Strainers

The strainers after the low-pressure pump remove material which might interfere with the action of high-pressure pump valves or plug the spray nozzles. The strainers consist either of wire mesh supported within a heavy wire frame or of drilled metal cylinders. In either case, the size of the openings is such that any material which is passed will also pass the spray nozzles. The total strainer area is large enough to allow a reasonable length of time between washouts or changes. Parallel units are provided so that one may be cleaned while the other is in use. Pressure gages on both sides of these units enable the operator to read a differential pressure and to arrange his cleaning schedule accordingly.

8.3.1.2.1 Mechanical Disintegrator

To eliminate frequent strainer changes, loss in material during strainer washouts, and frequent strainer replacement, a mechanical disintegrator to break up lumps is being used in some plants instead of strainers. The disintegrator operates very much like a hammer mill, having a wire screen basket within which a series of blades revolve.

Crutcher mix is pumped into the center of the basket and forced by the blades through the basket, breaking up any lumps in the mix.

8.3.1.2.2 High-Pressure Pump

The high-pressure pump is usually a triplex unit to minimize pressure fluctuation. Most units are capable of generating pressures up to about 200 kg/cm^2. The drive mechanism has a variable speed so that the rate at which crutcher mix is pumped may be readily changed to suit tower operating conditions. Water sprays on the pistons provide cooling, removal of crutcher mix, and packing lubrication. A high-pressure relief valve is also provided.

The feed line to the spray nozzles is usually traced and insulated to guard against makeup. Since this line is usually long, it must be made fairly large to minimize the pressure drop with the viscous crutcher mix. The line should not be left full of crutcher mix for more than a few minutes due to the danger of makeup.

The chief problems encountered in high-pressure pump operation are starving due to insufficient supply pressure and short packing life due to wear and scoring of pistons. To minimize this, pistons are usually hard chrome plated. A Chevron-type packing is used to withstand the high pressure.

8.3.1.3 Steam Injection System

A steam injection system is frequently installed between the drop tank and the low-pressure pump. A positive displacement rotary pump (e.g., Viking) is used to prevent steam from flashing back to the drop tank or the high-pressure pump.

Steam is injected through a series of nozzles into the short section of pipe following the pump. The steam condenses and heats up the crutcher mix thus improving atomization. Excessive steam injection can cause binding of the low-pressure pump; however, crutcher mix can usually be heated safely to about 110 °C if the rotary pump supplies a minimum pressure of kg/cm^2. Most factories which have these units often use them even when steam injection is not being used in order to smooth out the operation of the low-pressure pump.

8.3.1.4 Air Injection

Air is injected into the crutcher mix line immediately after the high-pressure pump. For best control, the air should be about 1–15 kg/cm^2 higher in pressure than the crutcher mix in the line. Air flow rates are low compared with crutcher mix rates and metering must be accurate to obtain good control. A sensitive needle valve is required. Makeup in the injection nozzle has been a problem especially after a shut down. A new nozzle has been designed which both meters and injects air. When the air is shut off, the nozzle is closed, preventing mix from setting up inside it. The air pressures required for injection are up to 70 kg/cm^2, thus requiring either heavy-duty air compressors or cylinder air.

In a few factories, air has been satisfactorily metered into crutcher mix after the Viking pump of a steam injection system. Air pressure requirements are reduced but control is not as good.

8.3.1.5 Spray Nozzles

The spray nozzles are located in a ring at the top of the tower and are fed from a common header. Steam blowouts are sometimes provided for cleaning or preheating the nozzles. The location, length, and configurations of the nozzle arms are such as to make the most effective use of tower cross-sectional area. Impingement of sprays on each other or on the tower walls must be avoided since agglomeration and excessive coarse production will result. Hollow cone pressure nozzles are normally used.

Aside from occasional instances of severe plugging, which are usually traceable to a defective strainer, the most serious nozzle problem is the marginally poor spraying which results from gradual wear of the insert. This is difficult to detect yet will produce excessive coarse scrap and poor finished product particle size. The nozzle inserts must be inspected frequently for wear, and any that show an imperfection (e.g., pitting and scratches), especially around the orifice, or in which the orifice shoulder has become rounded, should be discarded and replaced. Nozzles in use should be checked frequently, and any that do not spray a smooth, round cone should be cleaned. If the nozzle continues to spray poorly immediately after cleaning, the insert should be discarded and replaced. For best viewing of nozzles in use, they should be at eye level and well illuminated.

Crutcher mix lines and nozzles can be cleaned by flushing with steam or, preferably, with hot water. It is felt that steam can precipitate silicate and cause partial plugging of nozzles. After water-washing, the lines are left full of water during the shutdown and the water is then pushed out ahead of the crutcher mix at startup or allowed to drain to the sewer.

8.3.2 Air Flow

Outside air is heated in gas-fired furnace, mixed with additional outside air and delivered to the tower through ports just above the cone. The air enters the tower at about 20–40 °C. The inlet ports are not directed radially inward, but are partially turned in a tangential direction, with a slight downward angle. The air is thus given a helical motion or "swirl". In most towers, the degree of swirl can be varied by means of baffles placed in the inlet ports. Swirl is important since heat and mass transfer are directly related to air turbulence.

Air is removed from the top of the tower at approximately 90 °C. It passes next through the exhaust cyclone, where entrained fines or "overheads" are removed, then through the exhaust fan from which it is discharged to the atmosphere.

8.3.2.1 Furnace

The equipment used for heating the air for spray drying consists of a forced air fan and a furnace with automatic controls and temperature recording equipment. The combustion gases are mixed with bypassed air and introduced into the tower through a plenum chamber or special duct-work arrangement. Gas or fuel oil may be used. A relatively low grade (but low sulfur) oil may be used if suitable heating and fuel atomization equipment are provided.

The furnace temperature must be kept high for complete combustion. The temperature of the air stream to the tower is varied by changing the amount of fuel delivered to the furnace. The maximum possible amount of air is bypassed around the furnace, consistent with maintaining high furnace temperature and an adequate supply of air for combustion. The total amount of air going to the tower is usually fairly constant and near the maximum capacity of the exhaust air fan. The damping action of the forced air fan is also used to control the static pressure within the tower. A slight negative pressure is maintained in order to ensure that product-laden air does not escape from tower openings.

8.3.2.2 Inlet Air

The direction and velocity of the air streams delivered to the tower materially affect drying rates. The air is normally introduced at the top of the cone but in a downward direction and with sufficient angular velocity to form a strong, swirl pattern. This results in greatest turbulence, faster drying, lower exhaust air temperatures, and higher product density.

In the extreme case, however, high swirl can result in excessive wall plastering and agglomeration due to particles being thrown against the wall and each other while they are still wet enough to stick.

8.3.2.3 Spray Tower

Spray-drying towers vary in height from 11 to over 47 meters and in diameter from 3 to over 10 meters. The larger units are provided with a motor-driven brushing ring to aid in cleaning. In the smaller units, hand scraping is relied upon. Feeder valves were originally provided in most units to give a positive seal at the base of the tower. These were subject to makeup, however, and have since proven to be unnecessary if the tower static pressure is properly maintained.

8.3.2.4 Exhaust Air

The air takeoff at the top of the tower is through a large central duct. This can result in a high-speed central core of air in the tower. Baffles designed to form a one-foot annulus at the exhaust duct and thus correct this condition have been installed in the smaller towers.

The exhaust air system consists of a high efficiency cyclone, to remove entrained fine material, and an exhaust air fan. In addition, a temperature sensing element connected to the air temperature recorder is supplied. This in turn is connected to a high temperature alarm which shuts off the supply of fuel if a preset exhaust air temperature is exceeded.

The fine material which is removed in the cyclone may constitute as much as 8–10% of the gross product stream. It may be discharged to a scrap buggy or bin, or returned to the base of the tower where it may be mixed with the main product stream.

This latter procedure is commonly used only when a subsequent spray-on step is available to guard against excessive dust in the finished product.

Severe problems of cyclone and feeder-valve makeup can occur with sticky product. Cleanout and inspection schedules of reasonable frequency must be set up to guard against unscheduled shutdown.

8.4 CONTROL OF SPRAY DRYING

A considerable amount of instrumentation is available for the spray-drying process as aids to control. It is important that this equipment be kept in top condition and be referred to frequently if processing is to be controlled satisfactorily.

8.4.1 Suction Pressure at High-Pressure Pump

A pressure indicating gage is used to measure this pressure. The low-pressure pump speed is regulated to keep the suction pressure at the high-pressure pump within the 3–9 kg/cm^2 range necessary to prevent starving the high-pressure pump.

8.4.2 Crutcher Mix Temperature to Nozzles

This temperature is measured by an indicating thermometer at the high-pressure pump and is normally controlled by varying paste temperature. If steam injection (or a crutcher mix heat exchanger) is available, crutching temperature and crutcher mix temperature to nozzles can be independently controlled.

8.4.3 Paste Pressure at the Nozzles

Pressure indicating gages measure the paste pressure at the high-pressure pump and at the nozzles. High-pressure pump speed, air injection, paste temperature, and size and number of nozzles, all affect nozzle pressure. Pump speed, however, is the primary control.

8.4.4 High-Pressure Pump Speed

A tachometer at the high-pressure pump measures its speed. This instrument must be reliable if reproducible operation is to be obtained.

8.4.5 Injection Air Flow and Pressure

A pressure gage at the air nozzle measures the injection air pressure. Maximum pressure is limited by the pressure rating of the compressor, receiver, etc. Compressor controls should be adjusted to keep pressure fluctuations caused by compressors loading and unloading to 1 kg/cm^2 or less.

Injection air flow is usually controlled manually to a specific rotameter setting. A flow recorder-controller has been installed in some plants to provide more precise control.

8.4.6 Tower and Airlift Air Flows

Manometers and differential pressure gages measure static pressure and air flow in the tower and airlift systems. These flows, however, are relatively constant and require little attention.

8.4.7 Furnace Controls

A system of adjustable baffles is used to control flow of air through and around the furnace. A typical system is integrated as installed, but some with a separate combustion air fan to supply combustion air to the furnace; in these plants, the "splitter damper" has been welded closed.

8.4.8 Tower Temperatures

Inlet and exhaust air temperatures are presented on an indicator recorder located near the tower operator's normal station. Inlet air temperature is controlled by adjusting fuel flow to the furnace. Exhaust air temperature is a dependent variable and cannot be directly controlled.

8.4.9 Density

Many plants now use a continuous density indicator-recorder that measures the density of the product flowing into the buggies.

Proper control of finished product density is important since most of packing machines measure a certain volume of product and thereby attempt to add the correct weight of product to the carton. Bulk density at the time of packing is the important consideration, and experience is the only guide for determining the blown density necessary for correct packing density. Usually, an appreciable density increase will occur during storage with the change being greater for magnesium sulfate sprayed product.

Density variations of a cyclic nature usually correlate with the batch crutching cycle. Upsets which occur when a new batch is dropped from the crutcher to the drop tank can be minimized by not running the drop tank agitator, thus permitting each batch of crutcher mix to float on the top of the preceding batch. Another way of reducing these upsets is by leaving a larger tare in the crutcher. This effectively means reducing the crutcher batch size and increasing the number of batches made.

Recent experimental work has shown that the crutcher mix flow rate in the high-pressure line to the tower has a considerable cyclic variation, corresponding to observed density cycling. Flow rate changes are caused by changes in crutcher mix properties as the mix ages. To minimize these changes, a crutcher mix flow controller is being evaluated. A flow sensor transmits a change in flow to a controller on the high-pressure pump. This controller, in turn, changes the pump speed to return the flow to the original rate. Significant reductions in density fluctuations have been seen in short trials of this equipment.

Density upsets also frequently result from crutching variations. Poor paste moisture control, varying scrap usage, weighing errors, etc. will all cause poor tower operation. Uniform crutching is the most important factor in tower operation.

8.4.10 Product Moisture

The moisture content of the finished product is also extremely important since any deviation from the formula card tends to dilute or concentrate the product. Product moisture is quickly and easily checked with a moisture balance, however, and readily controlled by previously discussed techniques. The distillation moisture analysis provides a ready check on the moisture balance (e.g., Cenco).

A close correlation between product moisture and exhaust air temperature can generally be established. Normally, the higher the exhaust air temperature, the drier the product will be and vice versa. The only exception to this rule is that high product particle size distribution is conveniently plotted on the particle size paper. This paper shows the percentage of granules retained on a sieve as a function of the square root of the sieve size. A typical line for granules is free of break-up or agglomeration. A line displaced in the direction of coarser median particle size would result, for example, from a lower spraying pressure or from poorer atomizability.

Moistures can be obtained with high exhaust air temperatures when poor atomization occurs. An increase in exhaust temperature, along with an increase in coarse scrap rate, indicates poor atomization by one or more nozzles.

8.4.11 Particle Size

The granules obtained from the tower are not uniform but consist of particles with a wide range of sizes. Granule size is a combined result of atomization, puffing, agglomeration, and break-up. Atomizability of the crutcher mix determines the particle sizes obtained under given conditions of pump pressure, nozzle type and size, etc. As previously discussed, atomizability is a function of crutcher mix properties. By proper control or selection of crutcher mix properties, of pump pressure and of number and type of nozzles, it is possible, therefore, to control the particle size distribution at the nozzle.

It has been found experimentally that the distribution of droplets from the nozzle is not a normal Gaussian distribution, but that there are more large droplets than the normal distribution predicts. Therefore, a plot of particle size vs. probability does not give a straight line. However, if the square root of particle size is plotted on probability paper, a straight line usually does result. This "square root normal" distribution is strictly empirical with no theoretical basis. This distribution is typical of pressure nozzles and has been found to exist when spraying many different types of materials.

Since the factors causing break-up or agglomeration of droplets is external to the nozzle, any nozzle operating properly will produce essentially the same size distribution. This has been verified in extensive tests of nozzle types. Increasing the spraying pressure decreases the median particle size but does not change the distribution about the median. A change in atomizability of the mix also merely shifts the median size. Puffing increases the size, of course, and also affects the distribution somewhat since the very small and very large droplets do not puff, but this effect is small.

Agglomeration is the cohesion of two or more particles together to form a single larger particle. This causes an abnormal size distribution with a curvature at the larger sieve size end of the curve. Similarly, excessive break-up of the granules would result in a curvature at the smaller sieve size end of the curve. Such plots, determined for samples taken at various points in the process, can be useful, therefore, in helping to find the cause of excessive agglomeration or of excessive dustiness.

There has been some belief, recently, that a uniform, fine product may enjoy a marketing advantage. At the same time, however, increased tower rates have resulted in heavier tower loadings, more agglomeration, and coarser product. In addition, the amide and MgSO spray-on products used so widely in overseas brands also tend to be coarser. Consequently, control of particle size, based on the sieve test, is as important to product quality as control of density and moisture.

Some improvement can be made by changing formulation, such as adding STS to obtain better atomization. Most improvements, though, must be done by the factories themselves through careful testing. Some steps that can be taken are:

(a) Close attention to the number, size, and condition of the nozzles being used. Regular checks on the nozzles in operation and close inspection of the nozzles before use will prevent excessive coarse from too wide a spray angle or poor atomization.
(b) Optimum adjustment of the crutcher mix free moisture. For some formulas, the free moisture range is very critical.
(c) Optimum use of air injection. Air injection can be a useful tool for particle size control by enabling higher mix atomization pressures to be used.

The rate of coarse scrap production and the percent on 1-mesh are probably the best indicators of tower operation efficiency. Problems in crutching, atomization, and air flow within the tower are all reflected as excessive wet, coarse scrap.

8.4.11.1 Specks

Specks in the finished product can be a severe problem at times. These can arise from undispersed minor ingredients, burning or charring in the tower, poor atomization of perfume, or just plain dirt.

Note: Any reduction in sieve test can be taken as an equivalent reduction in granules density, if desired, by adjusting granules' particle size.
First step in curing a speck problem is to determine the source:

(a) An overall darkening of the product on the tower walls is not very common but can result from long periods of high exhaust air temperature.
(b) Accumulation of product in the vicinity of the air inlet will often burn due to contact with the very hot incoming air stream. In some units, this critical area is examined daily and scraped if necessary. In cases of burning, a thorough

scraping down or even a washout will at least temporarily reduce the incidence of burned specks but can, in turn, cause a problem with rust.

(c) The presence of rust specks in the product can be a serious problem. The addition of a stainless-steel liner to the tower is sometimes the only complete solution. Frequently, short occurrences of a burned or rust speck problem can be temporarily cured by deliberately blowing wet product for a few minutes and covering up, thereby, the sources of specks. A thorough inspection and cleanout is then made at the first opportunity.

8.4.12 Flow and Phosphate Reversion

Because of its effect on flow, STPP reversion must be minimized. Under normal conditions, most of the STPP reversion occurs in the tower when the water of hydration is driven off. Thus, for a given finished product moisture level, the STPP hydration in the crutcher must be optimized in order to minimize both reversion and stickiness (free moisture).

Individual samples of granules from various overseas factories range from 10% to 90% in STPP reversion. The average for these factories ranges from about 25% to about 60% reversion, of which 6% to 18% is present in the STPP receipt initially. Differences in reversion are caused primarily by processing conditions.

Reversion of STPP results in formation of more acidic phosphates which take off from the silicate and cause precipitation of silica as floc. The floc is insoluble in water and shows up in the wash solution and on poorly rinsed and dried articles.

8.4.13 Fragmentation

Dust is usually not fine granules but rather fragments of granules, resulting from break-up in the product handling system and during shipment to the trade. For most formulas, excessive dust indicates granules which are too fragile as a result of excessive over-drying and puffing. Some formulas (especially low active, low STPP, high sulfate formulas) appear to break up much more readily than others. The reasons for this weakness are not clearly understood, but experimental work is being done at present to increase knowledge of granule structure. The most recent work shows that substitution of about 10 parts of sodium carbonate for sulfate tends to strengthen granular structure.

8.4.14 Stickiness

Sticky granules can cause make-up of feeder valves, cause packing problems, and give a very poor product impression lumps and caking. Various formulas differ in stickiness. Severe stickiness can be due to amide formulas, which are stickier than non-amide formulas. For any given formula, however, the possible causes of excessive stickiness include:

(a) too high a temperature at time of packing.
(b) low sulfonation completeness.

(c) high free moisture.
(d) insufficient NaSO in the formula.
(e) accidental overuse of amide and some non-ionics.
(f) type of AB used, although this is generally rare in factory operation.

8.5 SAFETY

Synthetic granules are combustible and a danger of dust explosions exists. Towers are usually designed with excellent safety equipment. Systems and deluge units are built-in. Deliberate misuse of equipment or insufficient instruction of operators can cause difficulty. This should not be tolerated or encouraged in an effort to gain higher production rates.

8.6 CONCURRENT OPERATION

Most built towers operate counter-currently, i.e., air and particles flow in opposite directions. This procedure gives a fairly dense product since the droplets contact the very hot inlet air only after they have been partially dried and this reduces the amount of puffing which occurs. One of the chief disadvantages of countercurrent operation, however, is that the droplets are subjected to only moderate air temperatures initially. During this initial period, therefore, they remain fairly wet and sticky, so that contact with other granules or with the tower walls results in agglomeration. This also results in the production of quantities of coarse scrap and of a relatively large median particle size in the finished product. The fire hazard is also increased since the driest granules are exposed to the highest air temperatures.

In concurrent towers, the hot air is introduced at the top of the tower and flows downward with the product. Subjecting the droplets as they are formed to the high inlet air temperatures results in substantial initial puffing and lower finished product bulk density. However, the initial "skin-drying" effect of the hot air prevents agglomeration, so that the final granules are much finer and more spherical. With a suitable choice of process conditions, it has proved possible to make the normal range of densities. The major disadvantage of concurrent operation has been reduced production rates compared with countercurrent operation. Lower drying capacity is due to:

(a) Reduced evaporative driving force between the air and the particles.
(b) Reduced transfer coefficients due to lower air-particle relative velocity.

However, as many countries are moving to lighter density products, which are proving very difficult to make with countercurrent operation, concurrent operation may become more attractive.

8.7 OTHER METHODS

It is worth noting the synthetic detergent granules or powders may be made by methods other than spray drying. Spray drying, however, has proved to be the most economical way to produce a wide range of products in large volume with high quality.

8.7.1 Mechanical Mixing

Mechanical mixing has been, and still is, used by some manufacturers to formulate synthetic powders. Densities of dry mixes tend to be high and the range is narrow, raw materials are hard to find in suitable physical form, and segregation and dustiness are severe problems. However, when formulations, available raw materials, and product objectives permit, mechanical mixing may be the cheapest process. An example of such a product is Spic & Span, the household cleaner.

8.7.2 Agglomeration

Mixtures of small particles may be built up into larger ones by agglomerating them. Typical agglomerating agents are silicate and glue. The raw materials are pre-mixed and then agglomerated in a drum or mixer. Agglomeration increases particle size and reduces bulk density. By a suitable choice of raw material particle sizes and level of agglomerating agent, a wide range of densities and particle sizes may be obtained. Agglomerated products segregate less and are less dusty than mechanical mixes. Examples of agglomerated products are Cascade, for automatic dishwashers, and light density Spic & Span.

8.7.3 Dry Neutralization

Several processes that involve the spraying of acid mix onto a mixture of alkali in the presence of inorganic builders have been developed. In these processes, there is no drying step. One process uses powdered sodium carbonate (as the alkali). The acid mix is sprayed on in an inclined rotating drum and most of the moisture formed is absorbed by the STPP that is also present. Some of the moisture is removed by a stream of air. Range of formulations are limited by the product's water binding capacity since product moisture cannot be independently controlled.

9 Product Handling

Synthetic granules leaving the spray-drying tower are hot, contain oversize and undersize particles, and must be transported to a storage area to await packing. Additional ingredients, such as sodium perborate, magnesium sulfate solution, perfume, etc., are often added to the granules after they leave the tower but prior to packing. All these operations, cooling, screening, transporting and adding of ingredients, are performed in the product handling system which includes all the equipment from the spray tower to the finished product buggy. Thus, the product handling system consists of:

(a) spray-on drum
(b) product conveyor belt
(c) airlift
(d) product separator and fines filter
(e) coarse scrap grinder
(f) density recorder
(g) perfume box
(h) buggy filling station

Not all factories have these pieces of equipment. At the buggy filling station, the granules are collected in buggies or bins and sampled for analysis to ascertain that they meet finished product limits before being sent on to the packing department.

Each piece of equipment listed above performs one or more specific functions. These are discussed below under "Process Equipment". In addition, however, chemical and physical reactions take place within the granules as they pass through the product handling system. These reactions are discussed under "Temperature in Freshly Blown Granules".

9.1 PROCESS EQUIPMENT

9.1.1 Spray-On Drum

The spray-on drum is a rotating, inclined drum located at the base of the tower. Granules discharge from the tower feeder valve into the spray-on drum which, in turn, transports them to the belt. The drum itself is designed to provide:

DOI: 10.1201/9781003596219-10

(a) a means for obtaining maximum distribution of spray-on materials on the granules.
(b) a means for thoroughly mixing solid materials with the granules so that a uniform product is discharged from the drum.

When used for mixing solids, the spray-on drum is at least partly baffled to provide the best mixing possible. For spray-on, however, baffles are undesirable for two reasons. First, they lift the product too high, and there is then a tendency for product to build up on the nozzle tips. Second, they tend to expose the bare walls of the drum to the spray which may result in considerable build-up on the wetted walls. When used for spray-on, a light build-up will generally occur naturally in the drum. This prevents sliding of the product and provides adequate tumbling and mixing without the use of baffles.

The length of the drum will determine both the number of spray-on nozzles which can be used and also the degree of mixing obtained. In addition, however, the angle and speed of rotation at which the drum is operated are also important. High drum speeds are desirable since they tend to give a better mixing action. On the other hand, the greater the angle and the higher the speed, the less will be the amount of product present in the drum at any time. This can lead to inadequate coverage of the granules or to wetting of the walls of the drum by the spray-on material with resultant rapid build-up of product. In practice, none of these factors causes any problems and few difficulties are experienced with the drums other than an occasional tendency of the drive to slip.

The nozzles used for spray-on are selected to give a high degree of atomization. Both pressure and air atomization nozzles are used though the latter are generally preferred because of the better product coverage obtained. In both cases, the size of the nozzles is determined by the rate of spray-on required, the spraying pressure, and the number of nozzle positions available.

The number of nozzles is limited by the length of the drum since the spacing between nozzles must prevent excessive overlapping of spray cones. Normally, at least one idle nozzle or nozzle arm is kept available, even at maximum production rates, in case one of the other nozzles gets plugged or sprays improperly.

The distance of the nozzles from the granule bed and the direction of the spray must be adjusted such that the spray hits well within the boundaries of the tumbling bed of granules. Improper spray or direction of the spray may cause rapid build-up in the drum. The drum and nozzles should, therefore, be inspected frequently. Precautions, such as use of filters in the spray-on line, must also be taken to avoid plugging of nozzles. In the case of air atomization, the air pressure must be optimized so that adequate atomization is achieved without excessive mist formation. Mist will wet expose areas of the drum and cause gradual build-up of product.

9.1.2 Conveyor Belt

The only function of the conveyor belt is to transport the product from the tower or drum to the airlift. In some factories, the conveyor belt has been eliminated by discharging the product from the tower directly into the airlift. The capacity of the belt is a function of its speed of travel and of its width. The discharge of the conveyor belt is always some distance above the bottom of the airlift so that the air stream may attain a certain minimum velocity before reaching the product. Use of scraping knives, cleaning brushes, and special drive-pulleys minimizes the build-up of product on the belt itself.

9.1.3 Airlift

The airlift is an upright (generally vertical) duct through which air is drawn at a high enough velocity to lift product. The airlift fan is normally positioned after both the product separator and the fines filter.
The airlift serves three functions:

(a) *Cooling*. Product cooling, the primary function of the airlift, takes place mainly by direct contact of granules with cold air. Some cooling is also obtained by evaporation of water from the product. Thus, the degree of product cooling obtained in the airlift is a function of the temperature, humidity and volume of air used, and the free moisture in the product.

The volume of air drawn through the airlift is that required to provide adequate product cooling; warm product is sticky and will build up in the packing machines. With products containing heat-fused perborate, inadequate cooling may also cause "stacking" in the cartons.

(b) *Lifting*. The product must be lifted from the bottom of the tower to a height such that it can conveniently flow by gravity through the remaining processing steps and to the packing lines. Other methods, such as use of bucket elevators, may also be used but the airlift is preferred for its ease of operation as well as the fact that it cools the product as it lifts it.

While the volume of airlift air needed is determined by product cooling requirements, the airlift must be sized such that the velocity of the air is high enough to lift the product. However, it must not be so high that it causes undue breakage of granules with the formation of excessive fines.

(c) *Classification*. Classification of the granules by the airlift is carried out automatically since large, undesirable particles fall out of the air stream at the base of the airlift and very fine particles are carried through the product separator to the bag filter.

The material which drops from the airlift generally consists of large flakes or lumps of product such as fall off the tower walls, inlet air baffles, etc. These lumps are generally overdried and off-color from long exposure to hot air and contain large amounts of insoluble silica precipitated by phosphate reversion products. This part of the product cannot be added back in the crutcher with the normal oversize granules as dry scrap, but must be reworked, along with tower scrapings, in the wet scrap crutcher.

The amount and characteristics of this airlift scrap can be an indication of tower operation. Thus, if the particles are wet and lumpy, the tower is probably not drying properly. If charred or burnt specks are present, it is commonly assumed that there is burning occurring in the vicinity of the hot air inlet. If an excessive amount of good product drops, it is an indication of insufficient air flow in the air lift.

9.1.4 Product Separator

As the air stream leaves the airlift, it is drawn into either a gravity separator or a cyclone. Here, the product is separated from the airlift air. A gravity separator is a large settling chamber in which the air velocity is reduced sufficiently so that the product falls out of the air stream. The cyclone, on the other hand, is a conical or cylindrical chamber into which the air enters tangentially and in which the product is separated by centrifugal action. While ideally, maximum separation of product from air is desired, in practice, operation of the cyclone is designed to minimize granule breakdown so that less than complete separation is obtained. The carryover, however, is restricted to very small granules and granule fragments. With most products, the removal of these fine particles is desirable since it reduces the dustiness of the finished product.

Since the airlift fan is located after the product separator, the separator operates under a slight vacuum. For the separator to function, therefore, a seal must be provided at the product outlet. Either a feeder or flapper valve is used for this purpose. With products in which amide is added in the crutcher or which contain magnesium sulfate spray-on, build-up in either the base of the cyclone or on the feeder valve is prone to occur. An inspection and cleanout program of suitable frequency must, therefore, be established.

As the air stream leaves the cyclone it is drawn through a bag filter, which completely separates all remaining particles, and the clean air is discharged to the atmosphere. These filter fines are collected in a dry scrap bin and are reworked in the crutcher.

The fine particles tend to build up on the bag of the filter instead of dropping into the bin and periodic shaking of the bag is necessary. The fan must be stopped during this shaking to prevent the fine particles from being drawn into the fabric of the bag, making cleaning much more difficult. The severity of the build-up increases with increased moisture content of the product and, particularly, with increased humidity of the air. Normally, the bags require shaking only once or twice a day. Increased pressure drop across the bags indicates the need for shaking. In extreme cases of build-up, air velocity can be reduced somewhat so long as good product does not drop from the airlift. If shaking the bags does not clean them adequately, they must be removed periodically for manual cleaning.

Occasionally, difficulties arise with operation of the fan itself, such as, lipping of the drive or improper adjustment of the fan dampers. The air flow through the fan is relatively easy to check by removing one of the leads to the differential manometer and measuring the static pressure which the fan is pulling. Build-up on the blades of the fan can cause the fan to run out of balance and the noise generated indicates that something is wrong.

9.1.5 Product Sifter

To meet sieve test limits (e.g., "1-mesh" screen), the larger particle size granules must be removed. This is done in one or more product sifters arranged in parallel and located directly below the product separator. Either a vibrating-screen of the "Rotex" type or a parallel wire separator (PWS) is used.

Sifter is a unit which vibrates in a horizontal plane and screens the product through a conventional two-directional screen. The unit operates well with dry, crisp products but is inadequate for stickier granules, such as those with magnesium sulfate spray-on.

The parallel wire separator consists of a frame, mounted at a 45° angle, holding parallel wires, about 12 to the inch, in one direction only. As the granules slide down these wires, the good product goes through while the coarser particles go over the end of the frame into a coarse product chute. For best operation, careful adjustment of the distributor feeding the PWS is necessary in order to ensure uniform distribution of the product across all the wires. This is accomplished by means of a hinged gate within the separator feed hopper. Once the granules are on the wires, blinding of the screen can occur if the wires are not vibrated periodically. Originally, pluckers, which moved across the wires causing each to vibrate in a lateral plane, were used. However, sticky product tends to build up between the wires, ultimately damping the plucking action. With such products, slappers have been found to be more effective. These are rubber sheets attached to a rotating rod beneath the wires which slap the wires from below, causing them to vibrate in a vertical plane, at a frequency of about 200–300 beats per minute. The principal disadvantage of the slappers is that they generate a current of air which has a tendency to blow the product away from the wires, causing some carryover of good product into the coarse product chute.

The advantages of the parallel wire separator over the "Rotex" sifter are:

(a) it has higher capacity on both a cost and a space basis
(b) it has less tendency to "blind", particularly with sticky products such as those with MgSO spray-on
(c) it is easier to clean the wires than a screen
(d) it has lower power requirements.

Because of their greater effectiveness with products having MgSO, spray-on, parallel wire separators are now frequently used in plants making such products.

Both screen openings and openings between parallel wires are measured in "mesh", which is the number of wires per inch. The mesh size of the screens or parallel wires to be used is determined by the finished product sieve analysis specifications. The higher the mesh number, the smaller are the openings and the greater the quantity of coarse particles removed from the finished product. The capacity of a separator is limited by the quantity of "coarse" to be removed per hour. Accordingly, the capacity of a separator will decrease with increasing mesh number of the screen or parallel wires.

The capacity of a separator also varies with the crispness of the product. Because of the "wetness" or "sogginess" of products with MgSO spray-on, the capacity of the separators with these products is only about half their capacity when used with products not having MgSO spray-on.

9.1.6 Coarse Scrap Grinder

Coarse particles removed by the screen go to either:

(a) a buggy, from which they can be returned to the crutcher as dry scrap, or
(b) a grinding device, such as an Entoleter, which reduces the size of the coarse particles and then returns them to the main product stream in the airlift.

The Entoleter is a centrifugal, high impact grinder. The coarse particles are fed to the center of a horizontal, circular, high-speed rotor on which vertical pins are mounted. As the granules move out along the rotor under the influence of the centrifugal force, they strike the pins and the shroud and are thereby reduced in size. Some build-up tends to occur on the impact area of the shroud and this must be cleaned periodically. A canvas, rather than a metal shroud, is used since it can be cleaned while in operation by squeezing the flexible canvas material.

Grinding the coarse particles instead of reworking them as dry scrap increases production rate and may improve product quality by:

(a) reducing crutching time, thus eliminating the need for coarse scrap addition to the crutcher.
(b) reducing crutcher strainer and spray nozzle make-up.
(c) reducing air crutching, thus reducing the amount of light, coarse granules produced.
(d) increasing tower capacity, since additional water to absorb the scrap in the crutcher is not required.
(e) reducing the necessary coarse scrap buggy inventory.

9.1.7 Density Recorder

In some plants, the bulk density of the finished product is measured directly by the tower operator at the buggy filling station. A standard cup is filled with granules at the discharge stream from the duct into the buggy and then weighed. A refinement of this method is to let the sample of granules flow through a funnel set at a constant height above the cup, thereby minimizing the variations in the "packing" of the granules in the cup. The cup density is measured several times during the filling of each buggy.

To provide a continuous measurement of product density, however, many plants have now installed continuous density recorders. A portion of the main granules stream from the Rotex or PWS is diverted through a smaller sized supply tube into the density tube. The density meter is a vertical, plexiglass, scale-mounted tube which holds a constant volume of granules by filling up to the level of the discharge end of the supply tube. The flow of product through the tube is controlled by a constriction at its lower end. The weight of this fixed volume of granules is measured continuously to give an indication of the bulk density of the finished product. This bulk density is recorded in arbitrary units (which correlate with the product density as packed) on a continuous recorder at the density station where it can be observed by the tower operator and furnishes the basis for adjustment of the tower conditions to give the desired finished product density.

The operation of the continuous density recorder is sufficiently reliable that the cup density measurement is now used generally for calibration purposes only.

9.1.8 Perfume Box

The perfume box, in which perfume is sprayed onto the granules is located between the screen and the buggy filling station. Its operation is described in detail in a separate chapter.

9.1.9 Buggy or Bin Filling Station

Storage of granules in buggies or bins provides surge capacity between the granules making and packing departments. A standard high-buggy holds about 1,200 liters while the newer canvas bins hold 1,450 liters of finished product. Buggies are on wheels and are normally pushed by hand. Forklift trucks are used to move the larger canvas bins.

At the filling station, the finished product is diverted into one of two buggies or bins. Each buggy or bin is tagged with information on serial number, date, time made, deviation from standard density, and ultimate destination. Possible destinations are:

(a) *Pack.* If the granules meet the Finished Product Standards they are released for packing.
(b) *Blend.* If the granules have a minor deviation from the Finished Product Standards (e.g., low perfume content), the product is "blended", that is fed back into the airlift to be mixed with new product. The blending operation is controlled from the buggy filling station by the tower operator.
(c) *Scrap.* If the granules show a major deviation, such as dark specks, the product is normally scrapped, that is, returned to the crutcher and reprocessed.

While no operation is performed on the finished product in the buggies, many products undergo both physical and chemical changes during buggy storage. These changes, which are related principally to hydration of the phosphates in the product, can affect both the packability and the quality of the finished product.

9.2 TEMPERATURE RISE IN FRESHLY BLOWN PRODUCT

The conditions (i.e., temperature and time) under which granules are stored in buggies can greatly affect both the packability and the quality of the finished product.

A considerable amount of phosphate hydration can occur in granules during the first few hours after making. Product which has hydrated prior to packing will be crisp and free-flowing and will not cake in the cartons, whereas product in which a substantial degree of hydration has not been allowed to take place will be sticky, will make up in the packing machines, and will cake in the cartons. Since hydration is an exothermic reaction, the temperature of the granules will rise. If the product has not been cooled sufficiently in the airlift, hydration in the buggies may raise the temperature to the point where the granules become soft and sticky again. In addition, at high buggy temperatures (above 43 °C) perfume deterioration also becomes a problem. Thus, sufficient buggy storage capacity must be provided to ensure that the granules are adequately hydrated before packing. Also the product should be sufficiently cooled, such that it is never exposed to temperatures above 43 °C.

9.2.1 Phosphate Hydration and Temperature Rise in Granules

Granules at the base of the tower normally contain 5–10% moisture. In formulas using magnesium sulfate spray-on, this is increased to 13–15% H_2O in the spray-on drum. Part of the water is bound as water of crystallization in tripolyphosphate hexahydrate, the rest is present in the granules as free moisture.

The relative amounts of bound and free moisture in granules at the inlet to the airlift will depend on the degree of tripolyphosphate hydration in the crutcher mix, which varies with crutching conditions, and on the degree of tripolyphosphate reversion during spray drying. This will vary with both crutching and blowing conditions. The principal reversion product, tetrasodium pyrophosphate (Na,P,O), will generally be anhydrous at the base of the tower. In high active granules (35% ABS–35% STPP), approximately 60–75% of the residual (non-reverted) STPP will be hydrated.

Any phosphates, such as STPP and most of its reversion products, which are still un-hydrated at the density station, will tend to hydrate during storage in the buggies, causing a rise in granule temperature. Hydration, however, requires the presence of free moisture in the granules. Thus, the same factors, which control free moisture content, also affect the degree of hydration and the temperature rise of product in buggies.

(a) *Phosphate hydration in the product handling system.* The hydrate of pyrophosphate does not exist above 79.5 °C. Thus, the product in the tower normally will contain only anhydrous pyrophosphate. Spray-on of magnesium sulfate solution may, in some cases, lower the product temperature sufficiently to allow hydration to begin in the drum or on the belt. In any event, however, the shape of the hydration curves indicates that an appreciable amount of hydration probably takes place in the product handling system before the granules arrive in the buggies. This is probably particularly true for products with magnesium sulfate spray-on because of their higher free water content.

Base Granules Formula:

35% ABS
35% STPP

Approximately 12–14% H_2O normally is required to hydrate completely all constituents of 35% ABS–35% STPP formula granules. Thus, in products with MgSO, spray-on, enough water is normally available to hydrate all the hydratable material present. Any water in excess of the amount required to accomplish this will be held in the granule in solution in the AS phase. However, products without MgSO, spray-on contain only enough moisture (6.8%) to hydrate part of the hydratable material.

For ease of handling and to avoid caking in the cartons, more than about 60–70% of the maximum temperature rise of the granules should be allowed to occur prior to packing. Rates of hydration can be calculated from data, However, the lower the product moisture is, below about 12% H_2O, the more rapidly the point of maximum possible hydration will be reached for complete hydration. Values for 8% H_2O product do not represent equilibrium hydrations but only the degree of approach to the maximum hydration which is possible with a product moisture content of 8%.

The data illustrate why it is possible to pack granules with 8% H_2O soon after making, whereas with MgSO, spray-on product (15% HpO) it is usually necessary to let the granules hydrate in the buggies for about four hours.

(b) *Temperature rise*. Approximate values for temperature rise in granules can be calculated from the following data on heats of hydration:

Material:

Tripolyphosphate (Form II) 18.5 cals/mole Phosphate or 171 cal/gH,O
(Only Form II is formed in the spray-dried product.)
Pyrophosphate 24.3 cals/mole Phosphate or 135 cal/gH,O

Heat of Hydration. When calculated on the basis of parts water hydrated, the heat of hydration of SIPP is higher, but of the same magnitude as the pyrophosphate. The water binding capacity of pyrophosphate is, however, significantly higher than that of STPP:

Water binding capacity

Tripolyphosphate	29%
Pyrophosphate	68%

Acid pyrophosphates that do not form these high hydrates are also formed on reversion. As a general rule, therefore, the net water binding capacity of the reversion products will be approximately that of the tripolyphosphate itself. Calculations show that the temperature rise in the buggy is primarily dependent on the free moisture content of the product, i.e., the water that is available for hydration. It will, to a small extent, also decrease with increasing degree of phosphate reversion.

The calculated data check well with the experimentally determined data for the low moisture products. The high-moisture products, however, show much less temperature rise in the buggy than that calculated. The reason for this is that the easy availability of the water causes a substantial part of the hydration to take place in the product handling system, before the product arrives in the buggy.

The higher the moisture content the more hydration will take place in the product handling system. Therefore, it is often observed that low-moisture and high-moisture (MgO) products show the same temperature rise when measured in the buggy.

For both products, the temperature rise varies normally between 5 °C and 10 °C, the average rise being about 8 °C. This variation in temperature rise, as measured on the same product, is caused mostly by the degree of phosphate hydration in the crutcher and by factors such as over-drying and reversion of the product.

9.2.2 Evaporative Cooling in the Airlift

It has been found that granules lose part of their moisture in the airlift. This gives an evaporative cooling effect which can be fairly large. The amount of evaporation and cooling thus obtained will depend on product moisture, ambient air temperature and relative humidity, mass transfer coefficient in the granules, etc. Some typical values obtained in are:

Product moisture 31–32%
Percent moisture lost in airlift 15%
Product temperature at density station, 9 °C.

Everything else being equal, greater evaporative cooling will generally be obtained the higher the product moisture, as shown by the data above. In addition, where MgSO, spray-on is used, the spray-on itself will cool the granules in the drum. The combination of these two factors explains why products with 15% have lower temperatures at the density station than do products with 8% H_2O.

9.2.3 Control of Temperature Rise

In calculating or measuring thermal capacities of airlifts, four factors must be taken into account:

(a) sensible heat gained by the airlift air
(b) sensible heat lost by the granules in the airlift
(c) evaporative cooling of the granules in the airlift
(d) heat of hydration evolved by the granules in the airlift

From the standpoint of finished product quality, it is necessary that perfumed product not be exposed to temperatures above 43 °C. This means that the temperature of perfumed granules at the density station must be less than 43 °C by an amount equal to the expected temperature rise in the buggies. The temperature rise itself can be controlled to some extent by process conditions (e.g., by maximizing tripolyphosphate hydration). Nevertheless, in many warm areas, the ambient air characteristics are such that it is impossible to meet a specification of 3 °C maximum temperature in the buggies with normal airlift operation. In such cases, several different approaches are possible:

(a) Provide additional cooling of the granules by introducing ambient air into the cone of the tower or using a second airlift.
(b) Precool the airlift air (e.g., by refrigeration).
(c) Collect unperfumed granules, allow to hydrate and then pass the granules through a second airlift.

9.3 SCRAP HANDLING

Generally, scrap consists of granules whose physical characteristics are out of limits, but it may also include product, which is incorrect in composition. The aim of a scrap handling program is to work the scrap back into the process as quickly as possible and with a minimum of undesirable effects on process and finished product.

Scrap is reworked either through the crutcher or through a so-called "wet scrap" crutcher.

9.3.1 Wet Scrap Crutching

Scrap which is very old or material which has been scraped or has fallen from the tower cone or walls cannot be added directly to the crutcher. As a result of aging or of subjection to high temperatures for a long period of time such product becomes very hard and will not dissolve or break up in the crutcher. In addition, the SiO in the formula precipitates as a result of gradual acidification of the scrap, generally due to phosphate reversion.

The function of wet scrap crutching is to provide a long dissolving time for these dry, hard scraps and at the same time to resolubilize the silicate by treatment with additional caustic.

Many plants have a wet scrap crutcher. This is a tank provided with a screw-type agitator and with a means for heating the wet scrap by injection of live steam. It is common to process large batches of the scrap as they become available and then to use a constant small portion of the wet scrap slurry in each batch of crutcher mix. This is done to try to obtain uniformity of crutching and spray-drying by maintaining uniform scrap usage and crutcher mix composition.

In plants which do not have a wet scrap crutcher, wet scrap batches are processed periodically (e.g., once each week) in the regular crutcher. They are spray dried in the tower at the end of the day's operation and collected as dry scrap which is then added back to the crutcher in small quantities during regular operation.

9.3.2 Dry Scrap

Most of the scrap produced in the factories (e.g., coarse granules from the sifter and fines from the bag filters or exhaust air cyclone) is "dry scrap" which is worked back into the process directly through the crutcher. However, only a certain maximum quantity of dry scrap can usually be added to each crutcher.

The greater the quantity of scrap that is put back into the crutcher the greater the quantity which will be produced subsequently on spray drying. This is due to the fact that addition of scrap to the crutcher causes air entrainment. Furthermore, it is difficult to disperse dry scrap completely in the crutcher so that strainer and nozzle makeup and poor spraying tend to occur. For these reasons, efforts have been made to develop alternative procedures for handling dry scrap. Among such procedures are:

(a) *Fines return.* With products using a spray-on, the fines from the exhaust air cyclone may be returned to the spray-on drum and thus agglomerated with good product.
(b) *Coarse scrap grinding.* The coarse scrap stream from the sifters may be ground continuously and returned to the base of the airlift.

In reworking the remaining scrap through the crutcher, it should be noted that the scrap usage must be carefully leveled out. That is, each crutcher batch should be as similar to the preceding ones as possible. This is necessary to obtain uniform spray drying and to minimize the production of additional scrap.

9.4 INSPECTION AND CLEANOUT PROGRAMS

Adequate programs of inspection and cleanout must be established to protect finished product quality, to maintain efficiency of operation, and to prevent equipment damage from excessive build-up. Experience and reviews of operational and mechanical downtime records along with quality checks on the finished product will indicate which sections of the plant require the most attention. Programs of hourly, daily, weekly, monthly, and annual inspections and cleanouts can then be established. The greatest need for good cleanout procedures occurs generally in the tower and the product handling system.

9.4.1 Cleanout Procedures

Two methods can be used to clean out the tower and the product handling system:

(a) *Washing put with water.* This method usually is not too satisfactory for several reasons:
   (i) loss of the product which is washed away
   (ii) increase in equipment corrosion
   (iii) increased tendency for rust specks to occur in the finished product.
(b) *Scraping.* It is important that cleaning be thorough, with complete removal of all build-ups. After a thorough cleanout, product initially only builds up again slowly and there is a period of fairly clean operation. Once a layer of product has accumulated, however, build-up becomes rapid. Hence, it is important to scrape to bare metal.

A number of equipment modifications have been made in recent years to facilitate cleanout such as the addition of cleanout doors in critical areas and swing-out feeder valves for cyclones.

9.4.2 Tower Cleanout

The most critical area in the tower is the vicinity of the hot air inlets. If frequent cleanouts are not made here, the result is often burned or charred specks in the finished product.

The tower cone can become severely plastered with wet product if spraying is poor. Daily inspections and cleanouts, as necessary, are usually made. At least a weekly cleanout is standard for the tower cone.

The upper walls of the tower also often become severely plastered by wet sprays. This type of build-up tends to be self-limiting, however, and it is generally necessary to run the cleaning ring only weekly or monthly.

9.4.3 Cleanouts in the Product Handling System

The spray-on drum should not require frequent cleanout. Some build-up on the walls of the drum is desirable since it improves tumbling and mixing of the product. Generally, this is self-limiting. Difficulties are caused only if the sprays are misdirected and wet the walls of the drum. In this case, build-up may be rapid.

The airlift, product cyclone, and feeder valves must be inspected daily. The degree of cleanout required is dictated by experience.

Product sifters do not give much trouble. There is some build-up on the screens or wires but this is easily removed by brushing during operation. Hourly inspections are generally made.

The Entoleter coarse scrap grinder may have rapid and severe build-up, particularly with soft or sticky product. Build-up must be removed from the shroud two to three times per hour during operation. Hourly inspection of the rotor is time consuming and may be required as often as every two to eight hours. It is convenient, in order to minimize Entoleter downtime, to have a second rotor available to use while the first is being cleaned. The need for frequent cleanouts is the major current weakness of the Entoleter.

The perfume box should not cause difficulties unless the spray is misdirected and strikes the walls, although some build-up tends to occur at the product outlet from the box; one inspection per hour is typical, but the need for cleanout is rare.

10 Special Processing Steps

10.1 SODIUM PERBORATE ADDITION

Sodium perborate, an "oxygen" bleach, is used in detergents to improve the whiteness and cleaning of washed fabrics, mainly in localities where washing is done in very hot water or preferably, where laundry is actually boiled in detergent solution.

10.1.1 Chemistry of Perborate

10.1.1.1 Perborate Formula

The chemical structure of solid sodium perborates has three different hydrates that have been identified. Of the three, only the tetrahydrate is used commercially in detergent formulations.

Sodium perborate in aqueous solution behaves like a combination of hydrogen peroxide and sodium metaborate ($NaBO_5$), showing reactions characteristic of hydrogen peroxide. Thus, from a practical standpoint, sodium perborate acts merely as a convenient solid carrier for hydrogen peroxide.

10.1.1.2 Perborate Stability

Perborates vary considerably in their stability to loss of active oxygen during storage. Any loss of active oxygen during storage, either from the raw material or from finished product on the shelf, requires either increased usage of perborate in the product as made, thus increasing costs, or causes lower active O_2 content in the product as used, thus decreasing performance.

The liberation of oxygen from perborates is catalyzed by the presence of traces of heavy metal salts, such as those of iron, coppers, manganese, and cobalt, which are generally present in the perborate as received. Care should be taken that further contamination is not introduced by careless handling at the factory.

Although trace metal contaminants are fairly inactive as catalysts for the decomposition of dry perborate, they considerably accelerate its decomposition when free moisture is present. Dry perborates may, therefore, be relatively stable; however, decomposition may occur rapidly in solution.

DOI: 10.1201/9781003596219-11

The suppliers attempt to eliminate impurities from the perborate as completely as possible. They also add stabilizers, such as magnesium sulfate, to inactivate the residual trace heavy metals.

The decomposition of perborate increases rapidly with temperature. This is illustrated, for example, by the following data for a perborate of typical stability:

Temperature 20 °C
Active oxygen 0.2% per month

Perborate decomposition, therefore, is minimized by:

(i) Insuring a minimum level of heavy metals, and the presence of stabilizers in receipts.
(ii) Storing under cool, dry conditions.
(iii) Avoiding contamination with metals or metal salts.

10.1.1.3 Caking of Perborate Tetrahydrate Receipts

Basically, perborate caking occurs by one of two mechanisms, the first one being the more common one:

(i) Conversion of tetrahydrate to trihydrate. Above 15 °C, the trihydrate is the stable form and the tetrahydrate is only metastable Thus, at high temperatures, conversion of tetrahydrate to trihydrate may occur if seeds of trihydrate and free moisture are present. Caking results from rearrangement of the perborate crystal lattice and from the cementing action of the water released by the conversion. Thus, while some free moisture must be present in the tetrahydrate to initiate the reaction, the conversion is self-accelerating since water is liberated in the process.
(ii) Perborate decomposition. When perborate decomposes it loses active oxygen and metaborate is formed. The metaborate-perborate mixture obtained is extremely hygroscopic, absorbs moisture readily from the atmosphere, and causes caking.

10.1.1.4 Bleaching Mechanism

The bleaching action of perborates is due to direct oxidation-reduction reactions between hydrogen peroxide and the stains to be bleached. Oxygen liberated by the decomposition of the perborate is not effective for bleaching. Therefore, for maximum bleaching effectiveness, perborate decomposition both before and during use must be minimized.

The polyphosphate and silicate builders in synthetic detergent formulations are effective sequestrants for the trace metals always present in wash water and adequately stabilize the hydrogen peroxide against decomposition. Thus, in detergent solutions, even at the boil, perborate decomposition is not too rapid and provides real bleach effectiveness. Often, E.D.T.A. (sodium salt of ethylene diamine tetra acetic acid) is also added to provide additional sequestering and stabilizing power.

10.1.2 Methods of Perborate Addition

Perborate is added to base granules as a dry crystalline powder. The problems of perborate segregation and product caking require that the addition processes be carefully controlled.

Perborate segregation is defined as the variation in perborate level as found by the consumer between cartons, as well as between pours from the same carton. Part of this variation is caused by a combination of inaccurate metering and poor mixing; part is true "segregation", that is, de-mixing of previously well-mixed product. The degree of true segregation is a function of the differences in particle size and density between the perborate and the granules. The particle size of perborate can vary appreciably and it is, therefore, necessary to include strict particle size limits on perborate specifications.

One way of minimizing segregation is by fusing the perborate particles to the synthetic granules. The so-called heat fusion process consists of feeding the perborate and the hot granules into an inclined, rotating drum located beneath the tower cone. Since the tetrahydrate crystals melt at about 65 °C, a partial or total fusion of the perborate takes place. The perborate either melts on the surface and sticks to the synthetic granules or smears over the surface of the granules, depending on the degree of fusion. Upon cooling, the perborate resolidifies and adheres to the granules thus minimizing or eliminating segregation.

A potentially serious negative of heat-fused perborate products is the caking that can take place in the market. The caking can be caused by both moisture absorption and by high temperatures. The caking will get worse with increasing degree of fusion of the perborate. In hot climates, the heat-fusion process may give problems during processing also; inadequate cooling of the product in the airlift can cause serious packing problems and caking of the product in the warehouses. The trend, therefore, has been away from the heat fusion process and towards cold-mixing processes with their inherent segregation problems.

Cold-mixing of perborate can be done either before the buggy station or just before the packing machines. If the addition is done before the buggy station (in the airlift or after the product sifter), the perborate metering system can be tied to the tower operation so that a separate granules metering system is not necessary. Airlifting provides mixing, but a considerable loss of perborate in the airlift fines may occur. Adding the perborate after the product sifter eliminates the airlift loss, but requires a separate mixer (e.g., a cage mixer). In either case, some segregation can still take place in the filling and emptying of buggies. To eliminate this source of variation, some plants have resorted to adding perborate in a separate metering and mixing system just before the packing machines. The "Schenck" system and "unit mixing" are two processes now in use.

Both perborate and cold granules are metered on to separate weigh-belts into a baffled rotating drum (Schenck system) or into a small "cage mixer" rotating at high speed (unit mixing). In the Schenck system, the mixture is then transported by conveyor belts from the drum to the hoppers of one or more packing machines. In the unit mixing process, the cage mixer is located directly above the packing machine hopper, and one "unit mixer" supplies only one packing machine. Addition before the packing machine eliminates the segregation that takes place in the filling and emptying of buggies. The cost of these units is usually quite high since metering and mixing equipment is required for each packing machine (or each two machines in the case of the Schenck) and expensive electronic control systems are required. Of the two systems, unit mixing provides much greater operating flexibility than the Schenck and is easier to operate.

10.1.3 Processing Problems

10.1.3.1 Metering

Both the heat fusion process and the before-the-buggy cold mixing process are controlled by the active oxygen analysis of the product as it arrives at the density station. The problem is that of obtaining the right amount of active oxygen in the product. In both processes, the granules rate depends on high-pressure pump speed.

The perborate is metered either by a weigh-belt (e.g., Waytrol) or by a volumetric feeder (e.g., Vibra-screw). These units may be coupled to the high-pressure pump so that when the blowing rate is changed, the perborate addition rate is changed accordingly, or the control may be completely manual. The weight metering system of perborate is more accurate, because of the effect of density variations on the volumetric system.

The feeding rate of base granules may vary significantly, since:

(i) The relationship between high-pressure pump rate and actual feed rate to the tower is not exact because of variations in crutcher mix density, crutcher mix moisture, and occasional cavitation or starving of the high-pressure pump.
(ii) Even at constant feed rate to the tower, the product rate leaving the tower often cycles, probably due to slugs of powder falling off the tower walls and cone.

When starting up, the setting of the perborate metering system is adjusted to the high-pressure pump rate to give the right active oxygen content in the finished product. The product is analyzed for active oxygen every hour to check whether the relationship between high-pressure pump rate and perborate rate has changed. Product that is out of limits on oxygen content is normally blended in the airlift with freshly made product.

Both the Schenck system and the unit mixing system require frequent calibration of weigh-belts. Both systems are activated by the level in the packing machine hoppers. The rates should be adjusted as close as possible to the packing rate to minimize intermittent operation. This is difficult with the Schenck which, because of its high cost and its space requirements, must feed two packing machines. The complexity of its electronic control precludes the possibility of feeding more than two packing machines. These units are accurate within their design limits but require thorough maintenance to maintain accuracy.

10.1.3.2 Mixing and Segregation

Mixing is generally not a problem with the heat-fusion or airlift addition processes. When perborate is added after the product sifter, it is necessary to have a cage mixer ahead of the buggy.

In the Schenck system, the mixing drum should be baffled along its entire length to prevent segregation within the drum and the consequent variation in perborate level between packing lines as well as from carton to carton.

For sufficient mixing, the unit mixing system requires a cage mixer consisting of a small wire-mesh cylinder placed in a vertical duct and rotating around a horizontal axis. The mixing efficiency of this type of mixer depends largely on a sufficiently high velocity of rotation, as well as on the physical properties of the materials to be mixed. Experimentation is required to establish operating conditions for optimum mixing results.

A significant amount of segregation can occur in the packing machine hoppers. This segregation is minimized by keeping the hopper level low and as constant as possible. This, in turn, means adjusting the rate of the metering unit so that it is in operation as continuously as possible. A certain amount of segregation also occurs during the free fall of product from the packing machine flasks into the cartons. This segregation cannot easily be remedied except by reducing the product's built-in segregating tendency.

10.1.3.3 Sampling for Process Control

Sampling for process control becomes increasingly difficult as the product's segregating tendency increases. In the case of the Schenck and the unit mixing systems, whole cartons are sampled off the packing line and analyzed. In the other systems, sampling is done at the buggy station in as uniform and representative a manner as possible. Single grab samples are too strongly affected by segregation and short time cycling of product rate to provide a useful means of control. A composite sample, taken over one or more buggies, is required to establish perborate level in the product. In any case, a rigorous sampling technique must be established and adhered to in order to insure positive control of perborate addition level.

10.1.3.4 Scrap Handling of Perborate Products

The perborate in scrap that is returned to the crutcher is, for the most part, lost through decomposition in the crutcher mix. In addition to the economic loss of perborate, excessive evolution of oxygen may cause boiling over of the crutcher or drop tank and production of low-density granules by the same mechanism as air crutching.

To minimize these processing problems, addition to the crutcher of scrap which contains perborate should not exceed an amount corresponding to 5% active oxygen on a crutcher mix basis.

In view of these considerations, it is most important to blend out-of-limits granules rather than rework them in the crutcher. In fact, if the inventory of perborate-containing scrap becomes too large, it may be essential to dispose of it by means other than addition to the crutcher. Thus, for example:

(i) Packing room scrap which contains large amounts of good product can be blended in the airlift. The good product can be separated from the dust by installing a cyclone in the dust system prior to the dust filters. Coarse scrap (from the heat fusion process only) may be ground and blended in the airlift.
(ii) Filter dust may be agglomerated by a separate process and blended in the airlift.

10.2 MAGNESIUM SULFATE SPRAY-ON

The magnesium sulfate spray-on process protects against the humidity caking of synthetic granules that occurs when granules are stored under humid conditions in unprotected ("plain") cartons. A detailed discussion of caking mechanism and theories and the role of magnesium sulfate is given in later discussion. Briefly, humidity caking involves the hydration and inter-granular bridging of the inorganic salts on the surface of the granules. Magnesium sulfate inhibits caking by forming, on the granules' surface, magnesium phosphates which are not subject to hydration. About 2% MgSO solids on a finished product basis are required for good caking protection. The MgSO must be sprayed on as a water solution to be effective. Crutcher addition or dry mixing of solid MgSO does not provide the necessary surface coverage.

10.2.1 Process Description

Magnesium sulfate is received either as the anhydrous salt or as Epsom salts, $MgSO,7H_2O$, and is dissolved in water to a solution strength of 20–28% MgSO. Insoluble are removed by either settling or filtration depending on the quantity of impurities in the raw material. A strainer is located in the line ahead of the feed pump to remove any remaining particles that might plug the nozzles or cause excessive pump wear.

The solution is applied to the granules in an unbaffled rotating drum located beneath the cone of the tower. The operation of the spray-on drum was previously described. Either pressure nozzles or two-fluid nozzles can be used to atomize the solution. If pressure nozzles are used, the solution is normally delivered at a pressure of 0.5 kg/cm^2 by a Viking pump. Under these conditions, the pump can wear rapidly. Thus, recent experience suggests that air atomization at lower pressures is a preferred method. Granule coverage is equal or better, nozzle plugging is less severe since the two-fluid nozzles have larger orifices than the equivalent pressure nozzles, and pump wear is eliminated.

The rate of spray-on is controlled by a rotameter and is checked by comparing MgSO solution usage against production rate. The required concentration of the solution will depend on the moisture level of the base granules and the moisture loss in the airlift but is adjusted to give the formula quantity of MgSO (normally 2%) and approximately 13–15% total moisture. The moisture level in the base granules is normally maintained as high as possible. This minimizes agglomeration in the spray-on drum by reducing the amount of spray-on required. The solubility of MgSO (about 28%) limits the maximum base powder moisture to about 10%. Other advantages of maintaining high base powder moisture are increased tower capacity, reduced tri-poly reversion, and reduced temperature rise in the buggy. A minor disadvantage is the marginally poorer coverage of the granules surface by the more concentrated MgSO solution.

The moisture loss in the airlift depends on the temperature and relative humidity of the air and will vary, therefore, with the time of day and the season. On the other hand, in order to keep the amount of MgSO in the finished product constant, the rate of MgSO addition must be held constant. Thus, the finished product moisture must be controlled by careful adjustment of the base powder moisture. This is checked by analysis of samples of product taken from between the tower and the spray-on drum. In addition, during start-up, MgSO spray-on is not begun until the base moisture and other product characteristics are satisfactory.

10.2.2 Effect on Physical Characteristics

10.2.2.1 Density

MgSO solution spray-on causes a significant decrease in product density. Thus, in order to meet the same finished product density limits, products with MgSO must be blown to a higher base powder density than are non-MgSO products. This is another reason for maintaining a high base powder moisture, that is, to minimize steam puffing and to reduce the amount of water sprayed on.

10.2.2.2 Agglomeration

MgSO spray-on causes a certain degree of agglomeration of the granules in the drum. The total granule surface area available for absorption of spray-on solution decreases with increasing particle size. Hence, agglomeration will also increase with increasing particle size of the base powder because of increasing wetness of the granule surface. If the particle size is too large, the surface wetness becomes excessive and the product will "snowball", building up very rapidly in the drum. The successful operation of this process, therefore, requires a base powder of fairly fine particle size. Generally, sodium toluene sulfonate is added to the formula to aid in obtaining product of the required fineness.

10.2.2.3 Brightener Specks

If the MgSO solution is improperly applied, a problem of "brightener specks" can result. Unlike true brightener specks, which are due to poor dispersion of brightener, the MgSO-induced specks do not cause fabric spotting. A problem of brightener specks (when product is viewed under ultraviolet light) can be indicative of either poor brightener dispersion or poor MgSO spray-on. Samples before and after spray-on should isolate the cause.

10.3 SPECKLING

Speckled products normally consist of white product to which a small amount of colored granules, generally blue or green, is added. The amount of speckles added varies according to marketing requirements but is usually in the range of 3%. In general, finished product appearance is not critically affected by small variations in the level of speckles present; however, the particle size relationship between white base and colored speckles is very important to finished product appearance.

The amount of pigment contained in the speckles alone usually equals the total amount normally added to a uniformly colored product. This means that the concentration of pigment in each speckle will be from 10 to 30 times higher than in a uniformly colored product. This high concentration of pigment represents a potentially serious fabric staining hazard.

The major problems involved in preparing speckled products are generally those related to handling of the colored scrap. Contamination of white base product with this scrap will result in development of undesirable off-hues. Approximately 8 ppm of pigment in the base product is the maximum level that can be tolerated if optimum contrast between base product and speckles is to be maintained. This means that very little speckled scrap can be worked off in the crutcher. Several methods have been developed for making speckled product. The three processes currently used minimize the production of colored scrap as well as produce speckles of uniform particle size and appearance.

10.3.1 Crutcher-Added Speckles

This process is used primarily by factories regularly making a uniformly colored brand. The speckles are produced by spay drying. All the pigment required in the speckles is added in the crutcher, generally via a pre-dispersion tank. Tower conditions are adjusted to give a product which is somewhat coarser than average. Coarser colored particles tend to stand out more clearly as speckles, whereas very fine material (e.g., through 100-mesh) is not visible as individual speckles but merely gives a colored hue to the finished product. The screened, unperfumed prod is collected and stored until needed for blending into the white product.

The speckles can be metered into the main product stream either in the airlift or just above the perfume box. If the addition is above the perfume box, some metering device in the vertical product duct should be provided. In those plants using Schenck or unit mixing for perborate addition, the speckles can be added at the same time as the perborate. Since small variations in the amount of speckles added is not critical, accuracy of metering is usually not a problem.

To minimize the cleanout job required after spray drying a colored product, the following order of scheduling is recommended:

White product
Speckles for speckled product
Uniformly colored product
Speckled product (white base)

No color cleanout is required when going from white product to making speckles since contamination of speckles with white product does not matter. Only a minimal cleanout (scraping) is required when going from speckles to uniformly colored product. Again, no cleanout is required to change from a colored product to a speckled product since contamination of white base with colored product also does not matter. By the end of the speckled product run, the tower will have generally cleaned itself of most colored product so that the only cleanout required is a relatively easy one.

10.3.2 Color-Solution Spray-On

The spray-on method of making colored granules for speckled products is particularly attractive for factories not regularly making a colored brand. In this process, the colored granules are obtained by spraying the desired level of pigment onto white product. A spray-on drum is normally used for this purpose, but a cement mixer, a spray-on tube (such as those used for perfume spray-on), or a fluidized bed is also feasible.

The pigment itself is usually insoluble in water so that the pigment slurry tank must be agitated to prevent settling. The problems of coverage of the granules with pigment involve the same considerations of selection of suitable spray-on pressures, nozzles, dye concentration, etc. discussed in detail for magnesium sulfate spray-on. The quantity of water used in the slurry is determined by the degree of coverage of the granules desired. It is usually about 8% of the granules.

The colored granules may be produced either in a batch process or continuously. In the batch process, the granules are prepared by spraying color onto white granules and then storing the speckles until needed. As mentioned previously, coarse particles provide more effective speckles. Some factories, therefore, screen out a special, coarse fraction of granules for use as speckles. Blending of these colored granules with white base granules may be accomplished as previously described.

In the continuous process, the colored granules are obtained by diverting a portion of the screened white base product stream to a spray-on drum. Here, the desired level of pigment slurry is continuously sprayed-on and the freshly colored granules blended back into the main stream of white product at the perfume box. Some agglomeration of fine white particles with wet-colored ones dulls the brilliance of the colored granules. The reduced contrast between the speckles and the white base product gives a faded appearance to the finished product. Furthermore, coverage of the highly irregular granules' surface is far from complete and this also causes less intense and less even color than tower-blown speckles would have at the same pigment level. In many factories, the dyed granules are blended back at the airlift. This has the added disadvantage of introducing colored product into both the fine and the coarse scraps.

The blending operation should be designed to produce the minimum quantity of pigment-containing-scrap. Dustbins should be emptied before producing speckled product, and the addition of the speckles should not start until the base product is completely satisfactory. The contaminated scrap is then used either in colored brands or in white product at no more than specified levels. If too much contaminated scrap is produced and there is no colored brand in which it can be reworked, it may have to be discarded.

The spray-on method of producing colored granules has several advantages over blowing these granules in the tower:

(a) The tower is not contaminated with dyed products at any time, thus does not require cleanout between brand changeovers.
(b) Provision of extra buggies for storage of both colored speckles and colored scrap is not required.

The spray-on process also has the following disadvantages:

(a) It may lead to a more severe fabric spotting problem since all the dye is concentrated on the granule surface.

(b) It provides speckles which are less uniform than spray-dried granules.
(c) It provides speckles which are duller in appearance.

10.3.3 STPP Speckles

A third process for making colored speckles is the coloring of a coarse granular tripolyphosphate by a spray-on technique.

Speckles made exclusively from inorganic salts, such as STPP or soda ash, dissolve more rapidly than detergent-based speckles and, therefore, reduce the risk of fabric staining. Another advantage of this process is that it is completely separate and independent from the tower, thus increasing tower production rates, reducing scrap production and cleanout, and improving scheduling flexibility.

From the standpoint of appearance, STPP speckles are better (i.e., more evenly covered) than granule-based speckles made by the spray-on process, although not as good as tower-blown speckles.

If a sufficiently coarse grade of STPP is available, the process consists simply of a dye spray-on operation in a baffled rotating drum. However, if the available STPP is too fine to give sufficient definition of the colored speckles in the finished product, an agglomeration step to bring the particle size up to the desired level is needed. Agglomeration of a fine granular STPP is accomplished by spraying on 10–20 parts of dextrine glue solution on 100 parts of STPP, again in a baffled rotating drum. The process can be operated either batchwise or continuously. By controlling the glue moisture, glue builder, amount of spray-on, spray-on rates, etc., it is possible to control particle size and density of the finished speckles rather closely. As an example of desired particle size, approximately 80% of the speckles should be retained on a 35-mesh screen, since the through 5-mesh particles are too small to stand out clearly when mixed with the granules. Better dyeing efficiency is obtained when the pigment spray-on precedes the agglomeration step since the round STPP particles are easier to color than the irregularly shaped agglomerates.

10.4 PERFUME SPRAY-ON

Perfumes are important to the consumer acceptance of synthetic detergents. Not only are they pleasing in themselves, but they help cover up any base odor present in the granules. Unfortunately, perfumes are unstable. They are complex mixtures of volatile and reactive ingredients. Considerable care, therefore, is required in handling both perfumes and perfumed products to minimize odor degradation.

10.4.1 Stability of Perfumes

10.4.1.1 Temperature

Perfumes, both as is and when sprayed on product, are extremely sensitive to temperature. The reactivity of perfume constituents increases with temperature, and they can not only oxidize but may also react with each other or with the base product. In addition, there is also the possibility of distilling some of the more volatile constituents. In all these cases, degradation of the perfume occurs. Thus, there is a constant, gradual decrease in perfume quality as the temperature increases. Above about 43 °C, however, the rate of deterioration accelerates considerably. This is the maximum allowable temperature to which perfumes should be exposed at any time during blending, shipping, or storage.

The thermal stabilities of different perfumes as sprayed onto granules have been determined. Perfume odor quality drops significantly when either perfume or perfumed product is stored above 43 °C. Differences do exist between perfumes, however, in the quality of the degraded odor and in the rate and degree of degradation with time and increasing temperature. Thus, the relative sensitivities of these perfumes to temperature alone are:

10.4.l.2 Stability to Perborate

Perfumes vary in their stability to (or compatibility with) sodium perborate. In addition, perfumes are less stable with heat fused perborate than with cold mixed perborate. Most present perfumes are relatively stable at the 5% perborate level. At higher levels, however, differences between perfumes do show up as noted.

10.4.l.3 Stability on Aging

Perfumes may be fresh but deteriorate on aging. Stability is definitely hurt by 10% perborate.

Perfumes begin to degrade from the moment they are sprayed onto granules. The deterioration involves both chemical reactions and also loss of the perfume by evaporation. The stabilities of different perfumes to these two effects vary considerably, as already noted.

Perfumes also differ in the extent of their covering power, especially at lower levels of usage. A delicate perfume has relatively low covering power. As it evaporates from granules, not only is the perfume impression itself reduced but the base odor notes come through more strongly than they would in the presence of a more robust perfume. Protective cartons greatly reduce the loss of perfume by evaporation. Unfortunately, base odors are also retained within protective cartons together with the perfume thereby possibly nullifying any improvement in perfume quality. Certain perfumes which are suitable for use in cool climates may not be stable enough for use in hot areas where base odor development and evaporation and reaction of the perfume are all enhanced.

10.4.2 Processing

Perfume is sprayed onto cooled, screened granules either on a conveyor belt or in a perfume box or tube. A metering pump feeds the perfume from a gauge tank, through a rotameter, to the spray-on nozzle. The perfume is not turned on until all other product characteristics are satisfactory since any perfume in scrap is destroyed if it is reworked into the crutcher.

At normal usages, perfume is not visible on the granules but is evident only by its fragrance. It is adequate, therefore, to let the perfume spray hit only some of the granules. The perfumed granules will mix subsequently with the unperfumed ones in the ducts and during the packing operation and, in any event, the perfume migrates rapidly and uniformly throughout each carton of product. At higher concentrations, perfume will discolor granules, appearing usually as yellow or brown speckles.

Two basically different methods are presently used for spraying perfume onto granules. The first, in common use, consists of a series of pressure nozzles suspended over a traveling belt of product. The degree of coverage of the granules is a function of the spraying pressure and of the height of the nozzles above the belt. At low pressures, a coarse narrow spray is obtained which gives poor granule coverage. At high pressures, misting occurs which results in excessive loss of perfume. The second, more efficient method is to spray the perfume on cascading granules in an enlarged portion of the duct work located between the product screen and the density station. This inexpensive system is easy to maintain and results in better overall perfume coverage. It has found wide use in many plants in two slightly different modifications, the perfume box and the perfume tube.

10.4.2.1 Perfume Box

Product flows from the sifter into the perfume box at an angle of about 55° to the vertical, hits the opposite wall, and flows into the duct leading to the density station. The perfume is sprayed with a two-fluid nozzle located above the product stream. The nozzle is positioned as far from the product as possible consistent with having the perfume spray pattern within the boundaries of the product stream. This provides best perfume coverage with minimum build-up of product on the nozzle tip.

Air pressure is adjusted to the lowest pressure which will give good atomization. A coarse spray may cause perfume specks on the granules. On the other hand, excessive air pressure may cause loss of the more volatile components of the perfume with a resultant change in odor. It may also cause the perfume to penetrate the product stream and hit the wall behind or it may result in misting of the perfume which will then settle on and wet the walls of the box. In either case, granule dust will build up in time on the wet walls and fall off as large perfume-soaked lumps which appear as specks in the finished product.

For this reason also, it is important that perfume not be sprayed when there is no product flowing through the box, and that the flow of granules through the box be steady and uninterrupted. Perfume may also hit the wall if the product stream is of uneven thickness across its width. The evenness of the stream may often be improved by installing a hinged baffle in the duct ahead of the box.

10.4.2.2 Perfume Tube

In the perfume tube, perfume is atomized within a vertical, cylindrical curtain of granules, again with a two-fluid nozzle. The advantage of the duct over a box is that, with proper operation, there is no exposed wall on which perfume mist can settle and cause build-up.

The granules flow through a funnel onto a round horizontal plate. The granules build up into a cone on the plate and flow off in a cylindrical curtain. The location of the plate is adjusted to obtain an even thickness of the curtain around its periphery.

The nozzle is located immediately beneath the plate and its spray is directed downwards. To avoid build-up, there must not be any bend in the duct for several feet below the nozzle.

If build-up tends to occur below the nozzle, it can be prevented by installing an orifice plate below the nozzle. The product forms a natural funnel on the plate, and a short distance below the orifice, there will be very intimate contact between the granules and the perfume droplets. To prevent build-up on the underside of the orifice plate, the portion of the duct below the orifice plate is made slightly larger than the top part of the duct to form an annulus between the upper and lower sections. The slight vacuum formed by the falling granules will sweep the air through the annulus and across the bottom of the orifice plate and thus keep it clean of build-up.

The same precautions must be taken in the operation of the perfume tube as in the operation of the perfume box in order to secure good dispersion of perfume on the product and to prevent formation of perfume specks. These precautions are that:

(i) No perfume be sprayed when there is no product flow.
(ii) The thickness of the product stream be even.
(iii) The atomization air pressure be optimized.

10.5 TABLETED DETERGENTS

Tablets provide the consumer with a premeasured quantity of detergent in a convenient, compact form. They are designed to assure recommended usage and, thus, optimum performance. A typical tablet weighs approximately 2 oz. and is one inch thick and 2-1/2 inches in diameter.

10.5.1 Tablet Properties

A detergent tablet must disintegrate rapidly in use and yet be strong enough to withstand handling in the trade. These two properties are difficult to achieve without special processing techniques since they are in opposition. A strong tablet tends to disintegrate slowly; a rapidly disintegrating tablet tends to be too weak to withstand trade handling. A satisfactory tablet must also be resistant to surface abrasion and have a pleasing feel.

Strong, rapidly disintegrating tablets are achieved by pressing them from a mixture of spray-dried or flaked active and dry granular STPP. The tri-poly acts as an incompressible core about which the spray-dried granules are compressed. A network of continuous void space is produced within the tablet which permits rapid penetration of water.

A compacting pressure of 180–250 psi provides sufficient force to form fairly strong inter-particulate bonds without breaking down the wetting network. The strength of the tablet is usually increased by increasing the stickiness of the mix with a spray-on of water, nonionics, amides, silicate, glue, etc.

The pressed tablet is given a smooth, abrasion-resistant surface by spraying onto the tablet a small amount of water and then evaporating some of this water in an oven. The mechanism involves the solubilization of the material on the surface and its redeposition, together with the formation of interlocking STPP hexahydrate crystals.

10.5.2 Process Variables

10.5.2.1 Formulation

The product formulation has a large effect on tablet properties and also determines the cleaning performance of the product. Normally the dry granular STPP and the spray-dried granules are split about 50–50 in order to provide the proper wetting network. The adhesiveness of the mix is usually determined by the composition of the spray-dried portion. ABS-based granules are more plastic and adhesive than TAS-based granules which are brittle and non-sticky. By proper choice of spray-on and other conditions, however, it has been possible to incorporate both actives into a wide range of formulations. The mechanism by which water, amide, or nonionics spray-on increases tablet strength is believed to be different from that by which silicate or glue work. The former materials act like adhesive tape while the latter act as cementing agents.

10.5.2.2 Particle Size

Both portions of the tableting mix should be coarse enough to provide void spaces of sufficient size for rapid water penetration, but not so coarse that the tablet surface is rough. The 10-mesh and through 65-mesh sieve fractions should be negligible and 60–70% should be on 35-mesh.

10.5.2.3 Tablet Mix Density

The size and shape of the tablet is generally fixed by product appeal considerations. The weight of the tablet is then varied to give the desired performance properties and a good strength disintegration relationship. Increasing the weight of the tablet means a denser mix will be required to maintain the same strength disintegration relationship.

The spray-dried granules must be coarse, spherical, and dense so that a strong tablet can be made without excessive granule deformation and collapsing which would destroy the wetting network. Low crutcher mix moistures and crutcher mix deaeration with salt or ionics are necessary to avoid puffing during spray drying. Recently, the use of flaked, rather than spray-dried, actives has expanded the range of possible formulations (higher active), especially where cool water solubility is required.

Fortunately, the other major ingredient, granular STPP, is available in a range of densities from 0.5 to 1.1 grams/cc. The denser the STPP used, the more other ingredients can be added while maintaining a good strength-disintegration relationship.

10.5.3 Process Equipment

The heart of the tableting operation is the tablet press. The cost of this equipment varies tremendously depending on the capacity and degree of automation desired. No matter what the cost, however, the press must be capable of making tablets of uniform size and shape and must operate without having product adhere to the dies.

The active portion of the mix can be made either in a standard spray-drying tower or on a flake dryer. Mixing drums for spray-on and blending must also be provided. A tablet spray-on booth and oven dryer are necessary for the surface-curing operation previously described.

Packing equipment, like the tablet press, can vary from a simple hand line to a fully automatic line with polyethylene overwraps for the tablets.

11 Packing

Ideally, the goal of the packing operation is to place in the hands of the consumer a package of pleasing appearance, well-sealed and containing exactly the specified volume and weight of product. Package appearance is extremely important to consumer acceptance of a product. Thus, close quality control on incoming supplies on items, for example, such as the tendency of inks to smear or the ease with which cartons can be formed is imperative.

The packing operation itself includes the following steps:

(a) emptying buggies of product into the packing machine hoppers
(b) forming the carton or envelope
(c) adding a premium to the carton (in certain brands only)
(d) filling the carton or envelope with product
(e) sealing the carton or envelope
(f) packing the cartons or envelopes into containers
(g) sealing the containers
(h) palletizing the containers for transport to the warehouse

The packing room is normally located directly below the buggy storage floor. The buggies are dumped through the floor, and the product flows by gravity through a screen, into the packing machine hopper, the filling machine head, and the cartons.

In plants where perborate is added to the granules just before packing, the buggies are dumped into the feed hoppers of the mixing units, and the finished product either flows directly or is transported by belts to the packing machine hoppers.

11.1 BUGGY DUMPING

A buggy dump station consists essentially of one or two holes, located directly above the inlet to a packing machine screen, in the floor of the buggy storage area. There is one dump station for each packing line; the two holes allow two buggies to be dumped simultaneously so that the high-speed packing machines can operate continuously.

DOI: 10.1201/9781003596219-12

Buggy dumping can be a problem with products that have a tendency to crystallize or cake in the buggy. Removal of such products is facilitated by manually pushing the product out of the buggy by means of a pipe or pole.

11.2 SCREEN

The function of the screen is to break up or remove any lumps which may have formed during storage of the product in the buggy. Screening is particularly important, for example, with formulas containing magnesium sulfate spray-on. Rotex screens are generally used with carton fillers. Trans wrap machines are equipped either with simple, stationary screens or preferably with Rotex-type oscillating screens.

11.3 HOPPER

The packing machine hoppers provide surge capacity so that the packing machine itself can continue to operate while empty buggies are replaced by full ones. Most machine hoppers are furnished with an automatic level controller and either vibrators or an air injection system to prevent stacking of product in the hopper. With certain products, however, such as those containing cold mixed perborate, these conditions lead to excessive segregation in the hopper. Consequently, no air injection or vibration is used and instead, with these products, the level in the hopper merely is kept low to avoid stacking.

11.4 PACKING MACHINES

Synthetic granules are normally packed into either cartons, envelopes, or polyethylene bags by high-speed packing machines.

11.4.1 Carton Packing Machines

Several different makes of carton packing machines are used by various factories but all are similar in operating principle and all can run on different sizes of cartons. In most cases, cartons are received at the factory folded flat but with the side seam already glued. The cartons are opened as they are fed by belt into the machine. One end is sealed and the flaps are glued into place. Premiums, if desired, are added to the empty cartons, which then pass through the filling head.

Two types of filling head are used, weight fillers and volume fillers. With the weight fillers, a constant weight of product is discharged into each carton. Product from the packing machine hopper flows into a scale-mounted cup from which it discharges into the carton or else it flows directly into cartons while these are momentarily located on scales. With weight fillers, weight control is excellent but variations in product density will be reflected as variations in outage.

With volume fillers, a constant volume of product is discharged into each carton. The filling head consists of a rotating, horizontal, circular plate which has a series of flasks, or cylindrical cups, along its periphery. As the head rotates, the flasks pass beneath the product hopper and fill with granules. They discharge their contents into the cartons by opening a gate as the carton travels on the belt with the same speed and in the same path, for a short distance, as the flask. Different flask sizes are used for different carton sizes, but in addition the volume of the flasks can be adjusted slightly to get the correct weight of slightly of-density products into the carton.

The cartons are vibrated as they are filled so that they may be packed more readily in the short time available for the filling step. With these units, the outage remains constant, but if the product density is too high, too great a weight of product will be put into the cartons. Most of the newer machines used are of the volume filler type since these are able to operate at much higher speeds, and high speed and trouble-free operation are essential to economical packing.

The filled but still open cartons pass a photoelectric cell (electric eye) which discards any cartons which are not filled to the proper height. The tops of the acceptable cartons are then glued and folded into place, and the completely formed cartons are passed onto a compression belt in which the final sealing is accomplished. As the cartons leave the compression belt, they are packed into containers (cases), either by hand or automatically. The containers are then palletized and delivered to the warehouse by clamp lift (or forklift) trucks.

In the packing machines, there is a tendency for build-up of product to occur on all surfaces where the granules are exposed to pressure (e.g., at the gates of the flasks). This can be particularly severe when packing sticky or soft products (e.g., those containing amide or MgSO spray-on). Thus, dust control on machines is important since any small spillage of product tends to be smeared throughout the entire machine unless it is immediately sucked away in an airstream. One means of minimizing this build-up is to coat the critical parts of the packing machine with Teflon, to which granular detergents do not readily adhere. Nothing, however, substitutes for good inspection and cleanout of machines to prevent build-up and the resultant unscheduled stoppages. Evidence of poor clean-up is found when product builds up on guide rails and tends to mark cartons.

The carton forming steps are crucial, particularly with KB and wax-laminated board. Here, the quality of the packing materials is all-important. For example, if the cartons as received are not properly dimensioned, there will be difficulty in feeding them into the machines and improper gluing, folding, and sealing of the flaps will occur. Similarly, the type of glue which is used must be such that:

(i) it penetrates the cardboard of the carton
(ii) it is thin enough to be spread over the surfaces which are to be glued
(iii) it dries sufficiently quickly to form a good seal by the time the carton leaves the compression belt.

It is also important in carton forming that neither too much nor too little glue be used. If too much glue is used, it will be squeezed out over the ends of the cartons as they pass through the compression belt and will tend to glue one carton to another in the case. Too little glue, on the other hand, results in unsatisfactory sealing. Adequacy of the sealing is tested periodically by:

(i) taking two cartons and banging one against the other to see if product sifts from any of the seams
(ii) tearing open the cartons to determine the percentages of the flap areas which actually were glued.

11.4.2 Envelope and Bag Packing Machines

The Trans wrap, Triangle, and Hayssen machines are typical of those used for packing envelopes and bags. Envelope material is either a waxed or a polyethylene-coated paper, while bag material is most often thin film polyethylene. Both are received in single sheet rolls and formed into envelopes or bags in the packing machine itself. The first step in the operation is to form a long vertical tube by overlapping and heat sealing the edges of the sheet. The bottom of the tube is closed by another heat-seal. A measured volume of product discharges into the tube and collects above the bottom seal. A wide, horizontal heat-seal isolates the product from the rest of the tube; a knife cuts through the middle of the seal and separates the envelope from the remainder of the tube. The remaining half of the heat-seal now forms the new bottom of the tube; a new portion of product discharges into the tube, and the operation is repeated. This is known as Form-Fill-Seal machines.

The filling heads of these machines are similar to those of carton machines, except that the cups empty through a chute into the tube which forms the envelopes rather than into cartons.

The envelopes or bags are then transported by belt to a container or sack filling table where the units are manually counted into the container or sack. Great care must be exercised to ensure a proper count. This also maybe mechanized through automation.

11.5 WEIGHT AND OUTAGE CONTROL

Good weight and outage control depends on having uniform product (density and stickiness) to pack and also on carefully maintaining and adjusting the packing machine. Packing Standards specify the target weight for each unit, the maximum permitted variation in unit weight and also the maximum permitted outage. There is a difference, however, between Packing Standard limits and Control limits.

Packing Standards limits reflect the maximum variation that would be undetected by all but the most critical consumers. They also reflect any limitation imposed by governmental regulations. Control limits, on the other hand, reflect the normal variation that would be expected to occur due to random variation in the process. Assuming proper design and operation of equipment control limits are always much narrower than Packing Standards limits.

Control limits are used in conjunction with control charts to indicate when the measured deviations are larger than would be expected to occur by chance alone. This means that some action must be taken to determine and correct the cause of the deviation. Several good textbooks on how to set up and use control charts are available, and therefore, we will not present this material.

11.5.1 Coding

During packing, each carton and container or sack is stamped with a code number so that the product may be readily identified. The information given in a code will vary but it always includes the week (first two digits) and the year (third digit) in which the product was packed. Subsequent digits may represent the day of the week, the hour of the day and the line on which the unit was packed. Thus, the number 194 would represent the 19th week of 2024. Experimental products shipped to the trade are usually assigned a special code which cannot be confused with the regular code. Thus, coding permits identification of cases being held until they are released for shipment and also determination of the age of product in warehouses and on the market.

11.5.2 Premiums

Many of our brands now include premiums in the package. Such premiums give rise to numerous packing problems by complicating both weight and outage control, particularly if a random distribution of many different types of premiums is desired among the cartons in each container. Some premiums are low in volume and others high and this tends to cause variations in outage, even though the weight and density of the product itself do not change.

Premium addition can vary from just aiding premiums to the cartons by hand to the use of automatic devices synchronized with the high-speed packing machines.

12 Finished Product Quality

Consumer acceptance of synthetic granules is based on many quality items, some tangible and some intangible. One woman compares the whiteness of her sheets with those of her neighbor. Another takes delight in plunging her hands into billows of fluffy white suds. A third likes a dishwashing product to smell as though it would be mild to her hands. It is not easy to determine how customers judge competitive products and actually decide which one to buy. However, housewives do know what they like, even though they may not always know why they like it.

An important part of the technical effort is devoted to determining what housewives want and keeping up with them when they change their minds. Tests have been developed to measure these qualities in finished products. These include instrumental measurements as well as both actual and simulated use tests. They also include panel evaluations and blind tests (in which a subject is given a part of unidentified samples and asked to state a preference for one or the other, either in general or for a specific quality).

These tests and test results have been used to develop and improve synthetic granule products. They are, however, not feasible for use as control tests. Therefore, they have been translated into Formula Card Limits and Standards.

The quality items under consideration may be grouped as *inspection*, which includes color, odor, and lumps; physical characteristics, such as density, sieve, dustiness, and ease of pouring; *composition*, such as surfactant, ABS substitution (this is a measure of the split in surfactant between ABS and alkyl sulfate in mixed active formulas), phosphate, silicate, moisture, pH, and suds builder; and *performance*, such as detergency, sudsing, fabric whitening, etc.

A consideration of these quality items will serve as a summary of the previous sections. Process control is discussed as it affects each quality item. Troubleshooting checklists are provided, listing items in the approximate order they would logically be checked to determine the source of trouble.

DOI: 10.1201/9781003596219-13

12.1 INSPECTION ITEMS

12.1.1 Color

The principal cause of poor color in a product is dark paste. Therefore, checking paste color should aid considerably in determining the cause of any off-color product. Dark paste results primarily from charring in sulf(on)ation and to a lesser extent in neutralization. A high acid ratio may cause burning due to the high spent acid concentrations. Intermittent or non-uniform operation (as in start-up and shut-down) would have the same effect. Localized overheating and high spent acid concentrations resulting from poor mixing may be caused by wear in the sulf(on)ator or by a low recirculation rate. A high temperature spread is indicative of a low recirculation rate. A high temperature in any step of paste making can result in charring. In neutralization, erratic pH control may result in local areas of low pH and consequent discoloration due to accelerated corrosion of equipment. Dark paste may also result from impure acid or caustic.

Raw materials other than those used at the table may be responsible for color problems. In particular, the amide tends to darken at high storage temperature and may be dark either as received or as prepared for use. Minor ingredients, especially off-quality brightener, may also cause poor color.

Scrap is another source of poor color which may be very important. Excessive inter-brand contamination may be responsible for off-hues not readily traceable to their source. For example, excessive amounts of blue scrap will impart a greenish hue to a white product. Other scrap may also be dark, particularly that from the exhaust cyclone and airlift filter clean-outs. Re-use of such scrap should be spread out so as not to accentuate the darkening effect.

The importance of smoke-free combustion is basic. Presence of soot in granules will impart a gray or blue-gray color.

The above remarks have applied primarily to a uniform discoloration in a given sample. Specks of off-color product are more easily identified. Yellow and brown spots are usually a result of charring in the tower and indicate that a clean-out is needed. Colored spots indicate a non-uniform distribution of a dye or pigment.

Color Troubleshooting Checklist

Neutralized Paste Color

a. High spent acid strength (high acid usage). Start-up surges in acid.
b. Excessive sulfonation, sulfation, neutralization temperatures. Excessive acid mix hold time (especially batch units).
c. Poor circulation in acid mix dominant bath (vent gases from exchanger, check sulf(on)ator clearances, acid mix cooler temperatures).
d. Acid leaking into system ("O" rings in injection nozzles, vent valves).
e. Highly erratic or unusually high or low pH.
f. Color and quality of acid, fatty alcohol, AB, toluene, caustic. Poor circulation in paste system.
g. Sucking air into paste at caustic mix pump, paste pump (evidenced by air bubbles in surge tank, negative pressures at pump inlets).

Crutcher Mix Color

(1) Off-color scrap.
(2) Raw materials color and quality.
(3) Incorrect minor ingredient usage.

Blown Granule Color

a. Charring or specks caused by build-up at hot air inlets.
b. Poor combustion in primary air heater gives gray product due to the contamination with soot.
c. Poor atomization of builder, perfume spray-on.
d. Dark amide, perfume.
e. Rusty specks from agglomerates falling from upper tower wall.
f. Thin, hard-bluish flakes from feeder valves or product belt.

Perfuming is a major item affecting the odor of synthetic granules. It is important that the perfume be applied at the proper level, be well distributed, and not be overheated or otherwise degraded.

12.1.2 Odor

Base odor may be off due to sub-standard paste making, but this condition will nearly always be more obvious from color than from odor. In addition, spray drying can reduce the impact of base odors on finished products. Perfuming granules that are too hot can also cause loss and deterioration of perfume.

Poor odors have been encountered due to inter-brand contamination or impurities in raw materials. However, formulas, standards, and raw material specifications are now such that this condition should not occur.

Odor Troubleshooting Checklist

a. Perfume (proper perfume, well distributed, proper usage).
b. Poor paste odor.
c. High granules temperature.

12.1.3 Lumps

Lumps in the product are usually due to granules sticking together after they are made. They are formed primarily by compression, either under the weight of the product in a buggy or by compression in the discharge device of a filling machine. The tendency to form lumps is accentuated by having too high a moisture in the product for the degree of hydration of the phosphate or by having the product too hot as it goes to the bin. "Perfume lumps" are caused by poor spray-on of perfume and are recognized by their dark color and high perfume level.

Lumps Troubleshooting Checklist

a. Check sifters for holes, leaks.
b. If in packed product, check shaker screens at packing line hopper, product temperature, age of buggy (MgSO, product).
c. Check flowability of freshly blown product, moisture content, completeness, airlift temperatures, perborate usage.
d. Check perfume spray-on.

12.2 PHYSICAL CHARACTERISTICS

12.2.1 Density

Density is a fundamental and important characteristic of any granular product, since housewives measure products by volume. Product density and the percentage composition by weight govern the performance the housewife will get from products. As part of the fundamental product design, products research establishes the density required.

Several methods of determining density are used for different purposes. Although they are all similar, they may give different absolute values, and thus the method used and the point of sampling should always be stated.

(a) Densitometer density is used for measuring the density of packed product and forms the basis for Formula Card Limits and communications on question of density. It uses a specially designed instrument and gives precise results when properly used.
(a) Pour cup density and the continuous density meter are used for control of the tower in small and large plants, respectively. Appropriate density limits for these instruments must be determined at each factory for fresh product to ensure that weight and outage limits will be met at time of packing.

The packaging division designs the cartons on the basis of product density taking into consideration the amount of bulge to be expected.

The problem of density control involves the complex moisture-rate-density relationships discussed previously in spray drying. It is also recognized that sticky product, which may be light at time of packing, may increase in density (outage increase) as the product becomes crisper in the carton. This is usually due to a rearrangement of the particles and not due to granule breakup. So far, it has been impossible to state the exact result of changing the operating variables since so much depends on the formula being blown and the raw materials being used. The most probable causes of density variations may also cause problems with sieve-test.
Density Troubleshooting Checklist

a. Check exhaust air temperature (increase indicates poor atomization).
b. Check tower sprays, number and condition of nozzles, position of nozzle arms (sprays not overlapping or directed against wall of tower).
c. Check tower swirl.
d. Check nozzle pressure and high-pressure-pump speed.

e. Check low-pressure-pump pressure.
f. Check crutcher mix density, viscosity (crutcher amps), moisture, order of addition, scrap usage.
g. Check crutching temperature and crutcher mix temperature at nozzle.
h. Check air-injection rate.
i. Check paste moisture, density, temperature.
j. Check toluene or STS usage.
k. Check organic builder or MgSO, spray-on usages, and degree of atomization. Check MgSO concentration.
l. Check product dustiness caused by droplet fragmentation or granule breakup. Dust is usually very heavy.

12.2.2 Particle Size (Sieve-Test)

Particle size distribution is measured by the sieve-test. High sieve-tests results are usually the result of either poor atomization or agglomeration. Many of the factors affecting density also affect sieve-test. In addition, the items below should also be checked.

Sieve-Test Troubleshooting Checklist

a. See Density Troubleshooting Checklist.
b. Check for proper density (too light density increases sieve).
c. Check for agglomeration or large puffed granules.
d. Check sifters for holes, bad wires, and overloading.

Dust is primarily the result of granules breakage. Therefore, minimizing rough handling is an important first step in decreasing the dust level. In making granules, over-drying, which results in a granule which is more susceptible to breakage, is the primary cause of dustiness. Improving the distribution of the organic builder or MgSO, spray-on helps somewhat in alleviating dustiness.

12.2.3 Pour Grade

The fresh pour grade is dependent primarily upon the making conditions. The same conditions which promote lumping (i.e., too high free product moisture usually caused by too low phosphate hydration for the finished product moisture level and a high product temperature) promote this low pour grade or stacking. Poor distribution of organic builder or low completeness may also be responsible for a poor fresh pour grade.

12.2.4 Grit

A gritty or sandy feel of the finished product in solution is caused primarily by using STPP which is too low in moisture or is too coarse. Other sources of gritty, slow-dissolving material include:

a. Build-up of dry phosphate on the crutcher walls, which later breaks off in lumps.
b. Hard lumps in wet scrap.
c. Hard compressed lumps of dry scrap.

The problem of floc is the formation of insoluble silicate by removal of NaPO by reaction with either acidic phosphate reversion products or acidic paste. Over-drying in the tower is the principal cause of poor surface floc. To hold reversion to a minimum, the blown-product free moisture content should not drop below 1%.

12.3 COMPOSITION

In the final analysis, performance is essentially a matter of composition. The most obvious cause for an error in composition is a mistake in the actual amounts of the various ingredients which are put into the crutcher. Besides the possibility that a mistake may be made in weighing, it is also possible that the crutching formula as calculated may be in error. In particular, this could occur if the actual composition of the paste is not that which was assumed in calculating the crutcher formula.

Furthermore, a crutcher formula must be calculated for a certain finished product moisture. Deviation from this moisture will result, in effect, in diluting or concentrating the formula. Therefore, it is most important that the product moisture assumed in calculating a crutcher moisture be realistic and that it then be met.

12.3.1 Phosphate and Silicate

If either phosphate or silicate is out of limits in a finished product, it must be due to an error of the type. Neither phosphate nor SiO can be created or destroyed in making synthetic granules. Therefore, if the proper amount of these ingredients is put into the crutcher mix, the proper amount will be found in the finished granules by analysis. This does not, however, guarantee that phosphate or silicate will be in the proper form in the finished product. A substantial amount of the phosphate must be present as tripolyphosphate to be at the maximum cleaning efficiency. Silicate must be in the proper SiO_2:NaO ratio to avoid being precipitated as floc.

12.3.2 Moisture

As already mentioned, the composition of a finished product is completely dependent upon its having the proper moisture content. However, the form in which the water is present is also a consideration. Free moisture contributes to stickiness of the granules, whereas moisture which is bound as water of hydration does not. Control of STPP hydration is important, therefore, particularly with products which are inherently sticky or which have relatively high product moistures. Too low a moisture content must be avoided to prevent excessive floc and excessively fragile granules.
Moisture Troubleshooting Checklist

a. Check inlet and exhaust air temperatures.
b. Check rate, high-pressure-pump speed, number, and condition of nozzles.
c. Check low-pressure-pump pressure.
d. Check tower airflow.
e. Check paste moisture and temperature.
f. Check crutcher moisture and temperature.
g Check MgSO spray-on usage and concentration. (Check base powder moisture.)
h. Check perborate usage. (Check base powder moisture.)

12.3.3 Surfactant

Besides the general factors affecting composition, the total surfactant is particularly dependent upon paste composition. Having the correct paste composition depends upon having the proper paste moisture content, good completeness, and correct ratio of sodium sulfate to surfactant.

The correct ratio of STS to surfactant is also necessary since surfactant analyses must be corrected for the presence of STS.

The principal causes of low completeness are low acid ratio, weak acid, and incorrect temperature in sulf(on)ation. Too short a sulfonation time or too long a sulfation time, poor mixing due to corrosion in the sulf(on)ator or too low a recirculation rate, and intermittent or irregular table operation results in low completeness condition.

Low active to sulfate ratios are usually associated with improper operation of the acid mix concentration system, that is, wrong spent acid strength, incorrect settling temperature, too little or too much spent acid recycle, or air entrainment in cavitating pumps.

12.3.4 ABS Substitution

ABS substitution is a measure of the ratio of ABS to alkyl sulfate in mixed active formulas. Possible sources of error are incorrect table formulas, including a poor distribution of acid between ABS and AS systems, or failure to make adjustment for use of scrap from another brand.

12.3.5 pH

A low product pH is almost always attributable to a low pH phosphate. Phosphate which meets raw materials specification should not yield a low pH product. Hydrolysis of STPP in a crutcher mix and reversion in the tower tend to lower pH. However, these effects are not usually great enough to put a product out of limits for pH.

If such did occur, the product would inevitably be unsatisfactory because of silicate precipitation. Excess caustic added with treatment of poor scrap might conceivably give a high pH product.

12.3.6 Total Organic Builder

Total organic builder refers to the total content of amide and / or fatty alcohol in the product. An error in the level of spray-on builder results from a failure to balance the application of builder with the amount of product being made. The most likely cause would be failure to apply the builder at the rate calculated.

In the case of amide added to the crutcher, there is some loss in the tower. The amount of loss and decomposition increases with increasing tower temperatures and with decreasing finished product moistures.

12.3.7 Available Oxygen

Variations in available oxygen are almost always caused by metering variations or off quality raw materials. Precautions must, of course, be taken to protect raw materials and finished products from heat, moisture, and contamination with heavy metals or heavy metal salts.

12.4 PERFORMANCE

Most of the tests and analyses run by a factory on raw materials, intermediates, and finished products are designed to ensure uniformity in physical characteristics and in composition of each of the brands. However, each processing step and each formula is selected primarily on the basis of making a product which meets certain desired performance limits. The ultimate criteria of whether or not a given batch of finished product is satisfactory are normally (with the exception of appearance) the results of performance tests run by the Products Research Departments.

The following list includes only those items which might not show up in routine inspection and analyses.

Performance Troubleshooting Checklist

Performance Item	Possible Causes
Poor sudsing	Off quality AB, alcohol
Poor detergency	Poor brightener efficiency (seldom a problem)
Tarnish and corrosion inhibition	Amide, etc. (seldom a problem)
Severe sulf(on)ation conditions	Inter-brand contamination (low and high sudsers, TAS in all ABS formulas)
High unsulf(on)ated content	Excessive phosphate reversion
Off quality receipt (low activity)	Exposed to ultraviolet light (daylight and sunlight)
Pick up of moisture by brightener (therefore insufficient amount added)	Poor distribution in crutcher

13 Behavior of Product in the Market

In general, synthetic granules deteriorate with age in the market. The major changes that occur are loss of perfume, change in weight and outage due to loss or gain of moisture, and caking of the product normally associated with moisture pick up. Products containing perborate will also lose some of their available oxygen content. This discussion will be confined to the caking and outage changes that occur, the extent of these changes, and how they depend on product formulation and storage conditions. Many useful conclusions can be drawn by considering only two types of product formulations and three types of climate:

Products

a. Low moisture
b. High moisture (MgSO)

Climates

a. Dry
b. Cold-humid
c. Warm-humid

A high perborate formula, although it will analyze high in total moisture, is still considered a low-moisture product since the base granules are low in moisture.

The type of product (that is, high or low moisture) and the type of packing materials chosen depend on the climatic conditions of the marketing area. In practice, some marketing areas do not fit into any of the above classifications or have more than one type of climate depending on the season and geographical location.

13.1 CAKING AND OUTAGE CHANGES

13.1.1 Humidity Caking

When synthetic granules absorb moisture, they tend to form lumps due to the hydration of phosphates, particularly tetrasodium pyrophosphate (NaPO), on the surfaces

DOI: 10.1201/9781003596219-14

of the granules. This so-called "humidity caking" occurs whenever the relative humidity within the carton increases to a level at which a significant portion of the sodium pyrophosphate hydrates. Although pure NaPO hydrates at relative humidities between 50% and 60%, the synthetic granules will not cake until the relative humidity is above about 70%. This corresponds to an equilibrium moisture content of about 14% water for non-perborate products.

The relatively soft lumps formed at 70% relative humidity break up easily during normal product handling in the trade. However, the degree of caking very rapidly worsens with increasing humidity, and at the same time the lumps become much more difficult to break up.

This type of caking can be prevented, up to a humidity of 83%, by spraying onto the surface of the granules a solution of magnesium sulfate. The mechanism is believed to involve the formation of Mg phosphates on the surface of the granules.

These phosphates either do not hydrate or hydrate to a lower degree than the original phosphates, thus intergranular bridging is eliminated. Humidity caking of a MgSO product, however, does occur when the humidity is high enough to hydrate some other constituent, namely sodium sulfate, that hydrates above 83% relative humidity. At this point, ABS also begins to absorb a significant amount of moisture, the product becomes soft and sticky, and the granular structure is destroyed.

13.1.2 Stacking

This term is normally used for caking that can take place without moisture absorption, although in most cases it is aggravated by moisture absorption and high product moisture.

Stacking is most often caused by stickiness originating with the organic constituents of the product. The stacking tendency will, therefore, increase with the active level of the product. The tendency for stacking is greater with the more hygroscopic ABS products than with alkyl sulfate products. Also, straight-chain ABS is worse than branched-chain. The presence of amide always increases the stickiness of ABS products but the addition of STS can alleviate this problem somewhat.

Stickiness also increases with product temperature. Thus, product that is packed too hot often cakes in the warehouse. This caking may break up in normal handling and shipping.

Another type of stacking is found in heat-fused perborate products. Product packed too hot may contain molten perborate which causes caking in the carton when the perborate resolidifies on cooling. The problem is aggravated by the perborate decomposition that occurs due to excessive fusion. Such products may cake when exposed to high temperatures in the market, or overnight in the laboratory at 40 °C. At these temperatures, the partially decomposed perborate actually melts and gives a "demerara sugar" appearance to the product. Cooling the resolidified perborate cements the synthetic granules together and causes caking.

13.1.3 Carton Outage Variation

When synthetic granules absorb moisture, there is a decrease in product density. The product density decreases about 0.01 gm/cc for an increase in moisture level of 1%. This means that the granules will expand, if free to do so, about 4–5% for each percent increase in product moisture level and contract the same amount on moisture loss.

However, since the carton walls do not allow for expansion of the product on moisture absorption, some compression will take place, and the net outage decrease will not be as great. The extent of this compression depends on the way the carton is stored (flat, upright, etc.) and on the amount of handling it gets during the period of moisture absorption.

On moisture loss, however, there is no resistance to contraction or shrinkage, and the outage increase will be close to that indicated. The degree of moisture loss that is normally encountered is discussed below under "Equilibrium Moisture".

Products which undergo several alternating periods of absorption and loss of moisture can show, at times, a great increase in outage even if there is, finally, a net gain in moisture content. This is believed to be due to the compression that occurs each time the granules absorb moisture and expand.

13.2 MOISTURE ABSORPTION OF SYNTHETIC GRANULES

The moisture content of a product in the market depends on the relative humidity and temperature of the air, the packaging material used, and the length of time the product is stored. Given sufficient time, all products will reach an equilibrium moisture content which is a function of only the formulation and the relative humidity of the surrounding air. The rate of moisture absorption (or loss), however, depends on the air temperature and humidity, the packaging materials used, and the initial moisture content of the product.

13.2.1 Effect of Moisture

If an inorganic salt is stored in a closed system, it will exhibit a fixed vapor pressure of water which depends only on the moisture content of the compound and on the temperature. Note that the vapor pressure changes stepwise depending on which hydrate-pair are in equilibrium.

Increasing the temperature increases the vapor pressure exhibited by the hydrate-pair. The well-known Clapeyron equation (that helps us determine thermodynamic values for reactions or phases) describes the change in vapor pressure with temperature. If the data are plotted using relative humidity instead of vapor pressure, the temperature dependence is much reduced. An increase in temperature increases the vapor pressure of the hydrate, but the saturated vapor pressure of water also increases with temperature. Since relative humidity is defined as the ratio of actual vapor pressure to saturated vapor pressure, it does not change very much with changing temperature. Thus, for practical purposes, the equilibrium moisture of synthetic granules at constant relative humidity are not influenced by temperature.

If the salt is exposed to an atmosphere of fixed temperature and water vapor pressure, the salt will either absorb or lose moisture until the moisture content corresponds to the particular hydrate which is stable at that vapor pressure. Thus, the equilibrium moisture content corresponds to a specific hydrate. If several salts are present, each will absorb or lose moisture to form the hydrate that is stable at the given vapor pressure, and the equilibrium moisture content of the mixture will be the sum of the individual equilibrium moisture contents.

The data below give the equilibrium moisture of various salts as a function of relative humidity rather than vapor pressure.

13.2.1.1 Hydrates of Synthetic Granules Constituents

Most hydrates lose their moisture readily when exposed to humidities below that at which they form. Tripolyphosphate hexahydrate is an exception. It dehydrates very slowly and only at very low relative humidities. Furthermore, reversion of tri-poly occurs whenever the hexahydrate does lose moisture.

The equilibrium moisture content of the organic constituents in synthetic granules is continuous rather than stepwise. This is believed to be due to the fact that ABS, CNAS, amide, etc. are not one chemical compound but a mixture of various chain lengths and isomers. Actually, organics keep water in "solution" rather than as a hydrate. The equilibrium moisture relationship for ABS, Alkyl sulfates exhibit lower equilibrium moisture than ABS due to their greater crystallinity.

If the exact composition of the granules is known, particularly with respect to the phosphate reversion products, the equilibrium moisture can be calculated. It is easier, however, to determine equilibrium moistures experimentally. The equilibrium moisture of several different products data was obtained by exposing freshly made product (at formula card moisture) to an atmosphere of the given humidity and then measuring the product moisture after sufficient time had elapsed such that further changes in moisture could not be detected. Below 60% relative humidity, corresponded to about 1.2% below the formula card moisture since tripolyphosphate hexahydrate does not readily lose its moisture even when exposed to very low humidities. The moisture that is lost is considered to be "free moisture".

Experimental work has shown that the equilibrium moisture of many products is surprisingly close, especially if the sum of active and tri-poly contents is about equal. The explanation is that the equilibrium moisture of the active, STPP, and the reversion products are roughly the same in the practical humidity ranges.
There are, however, numerous exceptions:

(a) If the sum of the active and tri-poly contents is low and sodium sulfate is high (as in low-cost brands), the equilibrium moisture will be lower than normal.
(b) Perborate products will have a significantly higher moisture because of the high moisture of the perborate.
(c) Alkyl sulfate products will have lower equilibrium moisture, at high relative humidity, than ABS products.
(d) High moisture products (e.g., MgSO products) will have higher equilibrium moisture at low relative humidity than low moisture products due to the

stability of tripolyphosphate hexahydrate. They will also have lower equilibrium moisture at high relative humidity because of the hydration inhibiting effect of MgSO.

Above 80–85% relative humidity, the equilibrium moisture increases rapidly. This is the point at which sodium sulfate starts to hydrate. The active materials (ABS) also become increasingly hygroscopic. At 88% relative humidity, the equilibrium moisture may be as high as 30–50%. At this point, the product becomes so soft that the whole granule structure collapses.

1. 32% STPP » SO
2. 22% SIPP » NaSO
3. 17% SIPP » SO
4. 35% STP » MgSO

13.2.2 Rate of Moisture Absorption

The rate of moisture absorption is described by the basic rate equation:

Weight of water absorbed per unit time based on driving force and resistance. The driving force is the difference in vapor pressure outside and inside the package. The effect of temperature on the driving force can be determined by considering two special cases, i.e., constant relative humidity and constant absolute humidity.

In the first case, the driving force approximately doubles for each 10 °C increase in temperature. As can be seen from a psychrometric chart, the vapor pressure at high relative humidity increases much more rapidly with temperature than does the vapor pressure at lower relative humidity. Thus, even though the vapor pressure outside and inside the package both increase, the difference also increases. This effect is important when a product is being considered for two markets of the same relative humidity, but different temperature (e.g., warm humid vs. cold humid climates).

In the case of constant absolute humidity, the vapor pressure of the atmosphere outside the package by definition does not change. Inside the package, however, the vapor pressure increases with increasing temperature. Thus, the driving force for moisture absorption decreases. If the increase in temperature is great enough, the product will actually lose, rather than gain, moisture. This effect is important when trying to predict the behavior of product stored in a warehouse or store from humidity data obtained out of doors. Warehouse temperatures are generally slightly higher than outside temperatures and, therefore, the relative humidity is lower even though the absolute humidity is about the same. Retail stores are in general warmer still than warehouses. This is especially true in cold climates during the heating season.

The resistance to moisture absorption is mainly that of the packaging material. Generally speaking, plain board is very poor in moisture protection. KB board is significantly better, having 30 to 50 times higher resistance to moisture penetration. Wax-laminated board offers about 5 to 10 times more resistance than KB board. The actual resistance of the package is considerably lower since water vapor will pass through poorly sealed packages.

The tendency of water to penetrate carton board is measured as WVTR (water vapor transmission rate). The "resistance" is, generally speaking, inversely proportional to the WVTR value.

One other factor that must be considered is ventilation. Warehouses are generally not well ventilated and, therefore, the humidity changes much more slowly than it does out of doors. The effect of drafts (air flow) on the resistance is measurable but not important since the carton and container board resistances are orders of magnitude greater than air-film resistance.

The rate of moisture absorption is also very dependent on the physical arrangement of the stocks. Thus, cased products in palletized stacks have much less surface area exposed to the atmosphere than do individually stacked cartons. The moisture absorption rate is, therefore, much lower in warehouse product than in product undergoing rapid storage tests.

It is very dangerous to generalize, however, because the conditions may change so very much between stores and between warehouses. That is why products at the same age can show a very large spread in moisture contents when picked up from the market.

13.3 EFFECT OF CLIMATE ON CHOICE OF FORMULA AND PACKAGING MATERIAL

Both product formulation and packaging materials are selected to minimize caking and outage variations under the required marketing conditions. Weather and product age are both important in this regard.

13.3.1 Weather Conditions

Weather conditions in most areas of interest are summarized in the World Weather Data Handbook. The conditions in any given area will change during the day, from day to day, from season to season and even, for any one season, from year to year. Thus, to characterize a climate, the averages are taken over a period of several years of the temperatures and relative humidities of the three consecutive most dry and the three consecutive most humid months. For purposes of predicting caking performance and outage variations in existing or potential market areas, climates may be classified into three types based on the relative humidity during the humid season:

Range of average type of climate	Range of average temperatures	Relative humidities
Dry	20–30 °C	10–70%
Warm–humid	25–38 °C	70–83%
Cold–humid	0–20 °C	85–95%

(a) Dry climates (less than 70% average relative humidities throughout the year). Typical areas of this type include the inland areas of North Africa and the Near East and the highland portions of Mexico. In dry climates, synthetic granules do not cake even when equilibrium moisture is reached. They can, therefore, be packed in plain cartons. Products with low initial moisture (about 8%) will lose a negligible amount of moisture even at very low relative humidities.

(b) Warm humid climates (70–83% relative humidity) typically include tropical and subtropical areas. Examples include the Caribbean islands, the lowland areas of Mexico, Venezuela, most of the areas bordering the Mediterranean, and coastal portions of Morocco. In warm humid climates, low moisture products will require wax-laminated or KB cartons for caking protection. Since protective cartons only slow down the moisture absorption, enough moisture may be absorbed at the high temperatures of these climates to cause caking in 8 to 12 months.

MgSO products are now used extensively in warm humid climates, because they offer permanent protection, and because of the great cost savings in using plain cartons. The relative humidity in this type of climate very seldom drops below 70%, and the danger of moisture loss and outage increase is very low.

(c) Cold humid climates (85–95% relative humidity) include areas where the winter is very humid, but the summers may be relatively warm and dry. Examples include the northern European countries such as Belgium, Germany, the Netherlands, the Scandinavian countries, and the United Kingdom. At these very high humidities, MgSO is not effective in preventing caking and, therefore, protective cartons must be used for 100% protection.

Since the humid season does not last as long as in the warm humid areas, and since the rate of moisture absorption is lower at the lower temperatures, KB or wax-laminated cartons offer permanent protection. An exception to the use of protective cartons in these areas is if the product is to be stored in heated warehouses (and stores). In these cases, the indoor humidity is much lower than outdoors, and, therefore, the amount of moisture pickup is much reduced and plain cartons in bitumen containers, or even plain cartons in plain containers, may be satisfactory. If the product is to be stored in unheated warehouses, plain cartons in bitumen containers can be used if some caking can be accepted. The amount of caking will be much less than what occurs in warm humid climates since the rate of moisture absorption is much reduced at the lower temperatures.

(d) Mixed Climates. Many countries can have two distinctly different climatic regions. For example:

Area	Type of climate
Mexico: highlands lowlands	Dry / warm humid
Morocco: coastal areas inland areas	Warm humid / dry

Unless two different product formulations and packages can be justified, either a compromise formula must be adopted or inferior performance must be accepted. In Mexico, a 13% moisture MgSO product seems to give satisfactory caking protection in the lowlands without undue drying out in the highlands. In Morocco, the population of the inland areas is so small that no attempt is made to prevent the weight loss and shrinkage of the MgSO formula which is made principally for the coastal areas.

13.3.2 Product Age in the Market

The average age in the market depends on the turnover rate of the product. The age of the individual cartons, however, may vary from a few days to more than a year at the time it is consumed. This can also be true for brands that are fairly fast moving, although the percentage of old cartons will be much less than in a slower moving brand.

A brand packed in plain cartons may have good caking performance in humid climates if the majority of the cartons are bought before the product has had time to absorb sufficient moisture. A certain percentage of the cartons, however, will stay in the market long enough to cake. The percentage of caked cartons, therefore, is function of the average age of the product.

13.3.3 Prediction of Product Behavior

When going into a new market with an existing brand or old markets with a new brand, decisions have to be made on what type (i.e., low moisture or MgSO) of product and which packaging materials to select.
This decision is made on the basis of:

(a) Climatic conditions.
(b) Known behavior of similar brands under similar climatic conditions.
(c) Storage test results.
(d) Expected product age.
(e) Cost and availability of protective packaging.
(f) Degree of caking that can be tolerated.

The climatic conditions are judged from weather data information. The caking behavior of similar brands are based on market pickups in areas with similar climates.

Storage tests run on the product in advance of marketing are designed to give accelerated results. Satisfactory caking means that the product will be satisfactory in the market. A product that has unsatisfactory storage test caking, however, may still be satisfactory in the market because of the slower moisture absorption rate. In storage tests, therefore, it is important to run a product with known market behavior as a blank. The storage tests results will then indicate whether the new product is worse, equal to, or better than the known product.

The product age may vary considerably being significantly higher in the less developed areas (Africa, Asia, etc.) than in the U.S.A. or Europe.

The premium cost of protective cartons varies significantly from country to country. In some countries, protective cartons are not available at all and have to be imported at high cost.

If insufficient information is available to make a sound decision on product and packaging, it is normal to select low moisture products in protective cartons until adequate tests have been made to justify other alternatives. Ideally, no caking should be tolerated. Realistically, however, the importance of the market, the performance of competitive products, and the cost of protective packaging will determine how much caking can be accepted.

14 Material Balances

Material balances for synthetic detergent processes are relatively complicated, and a standard procedure has to be set up for calculating them so that the results will be uniform in all factories. A material balance is issued for each product and each formula card refers to a particular material balance. Material balances are particularly useful for accounting purposes.

A table of constants and assumptions, called "Synthetic Factors", used in calculating material balances is furnished to each factory. Every effort is made to see that the values used both in Synthetic Factors and in the material balances themselves are representative of conditions found in normal operation. Where variations exist, references are made to specific raw material specifications or to specific factory reports which report these variations.

The steps involved in calculating a material balance may be summarized as follows:

1. Use a specific form. Enter the desired finished product composition (including the miscellaneous gain) in the appropriate column. Usually, the relative amounts of sodium sulfate and of free AB or alcohol will be the only unknowns.
2. Calculate the amounts of AB, fatty alcohol, and toluene required to give the specified levels of ABS, AS, and STS, respectively, allowing for the expected sulfonation or sulfation completeness. Calculate the SO required for sulf(on) ating and the NaO required for neutralizing each of these.

In order to simplify the calculations, the acid is calculated as SO, and the water associated with it to form HSO is shown separately. Similarly, caustic and sodium carbonate are handled as NaO. All the water required to make paste of the required moisture content is shown to come in with the caustic or the carbonate solution.

3. When the unsulfated AB, alcohol, and toluene have been determined, allow for loss in spray drying (given in Synthetic Factors) and enter the remainder in the finished product composition.
4. Determine the amount of sulfate in the finished product by adding all of the known quantities and subtracting from 100%. Part or all of the sulfates will

DOI: 10.1201/9781003596219-15

be derived from the excess SO used in sulfonation. The possibilities are as follows:

(a) If the excess SO from sulfonation at minimum acid ratio is in excess of that required for NaSO, the acid mix concentration process must be used. Calculate the oleum required at minimum acid ratio, the amount of dilution water needed to reduce the spent acid concentration to 78%, and the amount of spent acid that must be removed.

(b) If the excess SO from sulfonation at minimum acid ratio is not sufficient to make all the NaSO, then several alternatives exist:
   - (i) Use an acid ratio within the allowable range which will provide the necessary SO for NaSO.
   - (ii) Return spent acid from storage if it is available from some formula that required acid mix concentration.
   - (iii) Put purchased NaSO into the neutralizer or crutcher.
   - (iv) Any economic combination of i, ii, and iii.

5. Calculate the Na required for the NaO. If NaOH is used for neutralization, calculate the split between carbonate and caustic required for the neutralization of sulfuric acid (see plant reports since it varies with factory and with the formula). Also calculate the amount of CO in the carbonate, the amount of CO retained in the product, and the amount of CO and moisture loss on the paste.
6. Determine the total solids in the paste and calculate how much water to add to give the desired paste moisture.
7. Calculate the crutcher formula entering both totals and a breakdown between solids and moisture contents. Since the amount of scrap to be used is not known, the crutcher moisture in the material balance is shown without scrap. For an actual formula, it will be easy to include the scrap and calculate the actual crutcher moisture.
8. Enter the amounts of spray-on or other materials added to the granules.

The following notes and calculations explain the material balance (see Appendix – Material balances):

1. Column 15 shows the finished product composition. Columns 1, 2, and 3 show the ingredients to sulfonation, and column 5 shows the ingredients to and from dilution and settling, which go to make the composition of the concentrated acid mix (column 6). Columns 6, 7, 8, 9, and 11 show the ingredients to neutralization; column 10 shows the CO and HO loss during neutralization and the CO remaining in the paste; column 12 corrects for water formed on neutralization; and column 13 shows the composition of the neutralized paste. Column 1 lists the losses that occur in the tower during spray drying.

   Rows A, B, C, D, E, F show the formation of finished product constituents ABS, free AB, STS, free toluene, NapSO, and Na_2CO_3, respectively. Rows K, L, M, N show the ingredients added during crutching. Rows R, S show the MgSO spray-on.

2. The finished product composition for the following constituents is fixed by performance or processing reasons.
3. To determine CMC usage, we must determine the purity of the CMC that will be used. This information is on the RMS. Assume that CMC purity is 66%. The CMG usage is 0.225/0.66 = 0.3+ parts. Enter this figure in column 15, row N.
4. Enter the other figures given in Step 2 in column 15. Include miscellaneous gain in spray drying for which carbonate neutralization is 0.20 (Synthetics Factors).
5. Working from right to left, line A is filled in by calculating the parts of AB required to form the indicated amount of ABS.
6. Since the acid mix concentration process will be used, we can arbitrarily fix the acid ratio at the minimum necessary for good completeness. From experience we know that acid ratio with 30% oleum will give 98% completeness:

$$\text{Total oleum} = \text{Total AB} \times \text{Acid Ratio}$$

7. The formation of toluene sulfonate and free toluene (rows G, D) is calculated in the same way as described above for ABS and free AB, using the appropriate completeness and tower loss figures. The SO remaining in the acid mix can now be calculated by difference.
8. The amount of CO remaining in the paste is known from analyses of past production of similar paste. In this example, it is assumed to be 0.2l parts, which is equivalent to 0.50 parts of NaPO in the finished product.
9. The sodium sulfate content (7.00 parts) in the finished product can now be calculated by difference (column 15). In this example, the NaSO will come from the neutralization of some of the excess SO from sulfonation.
10. The dilution water needed to reduce the spent acid in acid mix to 78% HSO is calculated from the definition of spent acid strength.
11. The composition of the concentrated acid mix depends on the ratio of STS to ABS. Synthetics Factors gives the relationship.
12. The spent acid removed in dilution and settling (column 5) is calculated as the difference between the concentrated acid mix composition (column 6) and the free SO_3 and HPO contents of the diluted acid mix before settling (column 3).
13. The spent acid returned to neutralization (column 7) is calculated as the difference between the SO needed for sodium sulfate and the one provided by the concentrated acid mix.
14. The NaO required for neutralization is split up as follows:
    (a) A1l of the NaO for the ABS and STS comes from NaCO.
    (b) All of the NaO to neutralize the CO left in the paste comes from caustic soda (column 11).
    (c) 96% of the NaSO for sodium sulfate is supplied by the NaSCO and the balance from caustic soda (Synthetics Factors).
15. The moisture content of the paste (column 12) is specified by processing conditions. In this case, it is assumed that 50% HO is desired.
16. The water under the Paste Loss column (column 10), 1.22, is the water lost to the atmosphere as vapor, as calculated according to Synthetic Factors (27% of

CO, loss). The water under the NaCO column (columns 8 and 9), 4.81, is the remaining water that needs to be added to give the calculated paste moisture. The total figure under the NaCO columns, 56.07, is the total weight of sodium carbonate solution added. The water in the carbonate solution will have to be corrected in the factory for the water that comes in with the caustic soda.

17. Columns 16, 17, and 18 describe the crutcher mix composition. After spray drying, the crutcher mix gives the finished product shown under column 15. Columns 17 and 18 break down the mix into solids and water content. Note that the crutcher mix composition indicated is for a mix with no dry or wet scrap added.
18. Rows R, S show MgSO solution spray-on. Synthetic Factors give the moisture loss from the granules in the air lift.

Appendices

Appendix 1 Synthetic Granules Process

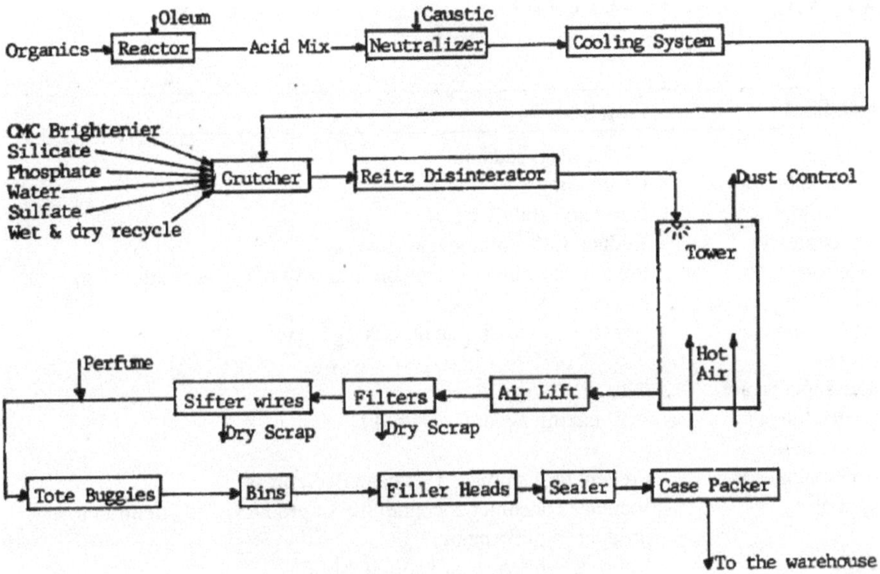

Raw Materials Used: Alkyl benzene, oleum, caustic, phosphates, sulfate, silicate, brightenier, perfume.

Appendix 2 Synthetic Detergents Ingredients and Function

Ingredient	Function
Surfactant	Detergency, sudsing
Inorganic builder	Detergency
Suds builder	Build and stabilize suds
Suds depressor	Reduce suds volume and stability
Silicate solids	Prevent aluminum corrosion. Improve physical characteristics of product
Sodium toluene sulfonate	Improve processing. Anti-caking agent
Sodium perborate	Bleach
Magnesium sulfate	Anti-caking agent for humidity caking
Brightener	
Tarnish inhibitor	Prevent tarnishing of German silver, brass
Pigment	Increase consumer acceptability of product. Call attention to a product improvement
Dye	Improve whiteness of washed fabric
Perfume	Increase consumer acceptability of product. Cover the base odor of the product
Bacteriostat	Prevent odor development on fabric or in wash water
Sodium sulfate	Diluent brought in during processing. Cheapen formula
Borax, sodium carbonate	Cheapen formula (carbonate reduces dust)
Water	Cheapen formula, prevent dustiness, anti-caking agent

Appendix 3 Surfactants

<u>Classification of Surfactants</u>

The organic active materials used in synthetic detergents are surface active agents (surfactants) with a high cleaning ability. Chemically, they contain a long, water insoluble (hydrophobic) chain, usually hydrocarbon, with a water soluble (hydrophilic) group at one end. Surfactants are commonly classified as follows:

(a) <u>Ionic</u>

Compounds of this type ionize in water. They are subdivided in three classes:

(i) <u>Anionic</u>: The active portion of the molecule, containing the hydrophobic chain and hydrophilic group, is negatively charged in solution.

Examples:

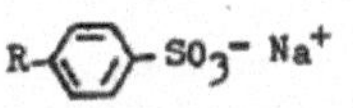

(Alkylbenzene Sulfonate)

$R\text{-}O\text{-}SO_3^-$ Na^+ (Alkyl Sulfate)

$R\text{-}COO^-$ Na^+ (Soap)

(ii) <u>Cationic</u>: The active portion of the molecule, containing the hydrophobic chain and hydrophilic group, is positively charged in solution.

Example: $R\ (CH_3)_3\ N^+\ Cl^-$

These compounds are used primarily in detergents for germicidal and disinfectant applications. Cationics are poorer in detergency, less mild, and usually more expensive than anionics.

(iii) <u>Ampholytic (Amphoteric)</u>: The active portion of the molecule can be cationic or anionic, depending on the pH of its solution.

Example: $R\text{-}\overset{H}{\underset{H}{N^+}}\text{-}CH_2\text{-}COOH \underset{H^+}{\overset{OH^-}{\rightleftharpoons}} R\text{-}\overset{H}{N}\text{-}CH_2\text{-}COO^-$

At present, only a few compounds of this type are available and their performance characteristics are still largely unknown.

(b) <u>Nonionics</u>

These compounds do not ionize in water. The most widely used materials of this class are made by addition of one or more ethylene oxide and/or propylene oxide groups to the hydroxyl group of alkyl phenols or aliphatic alcohols. The ether group, resulting from the addition, forms the hydrophilic part of the molecule. Many varieties of these compounds are available, but they are generally less effective in cotton detergency than anionics and generally low sudsing. Their primary use in detergent granules is as a suds depressor for low sudsing products.

Appendix 4 Chemistry of Sulfonation

CHEMISTRY OF SULFONATION

Sulfonation Reaction

The desired sulfonation reaction is:

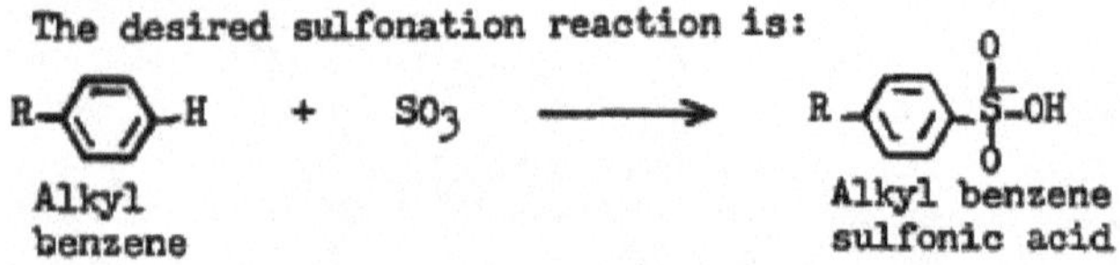

R = C_9 to C_{15}

The reaction takes place at an active hydrogen, primarily in the para position and the product has a direct C - S linkage. The reaction is not reversible.

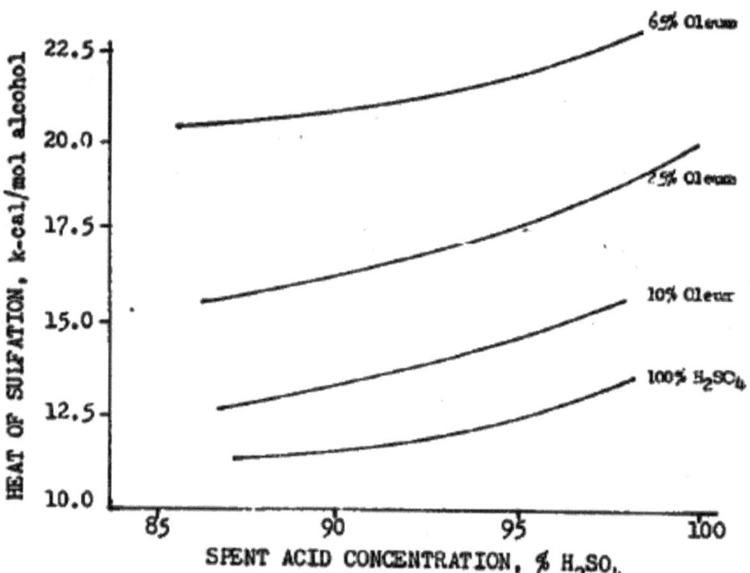

Net Heat of Sulfation of Coconut Alcohol

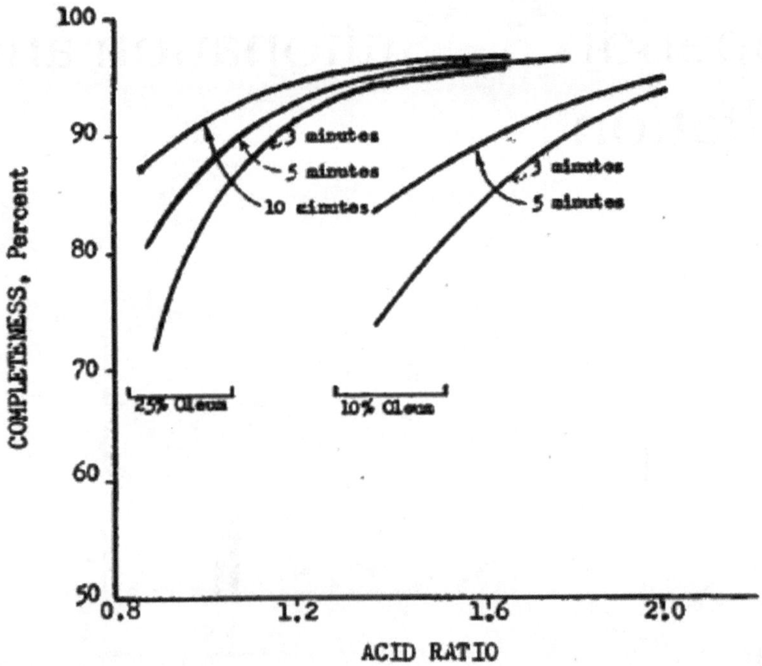

Completeness of Alkyl Benzene Sulfonation

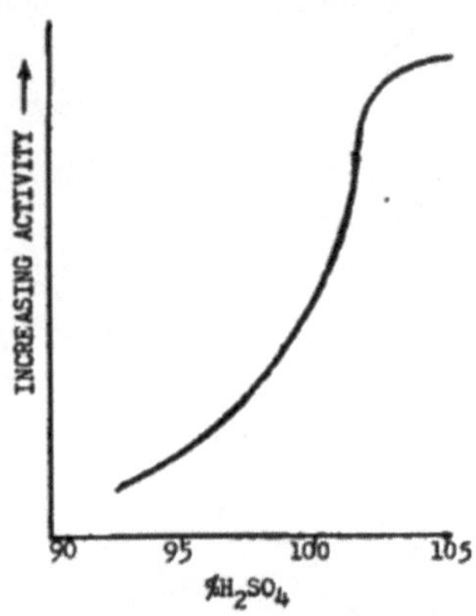

Activity of Sulfuric Acid

Appendix 5 Sulfonation and Sulfation

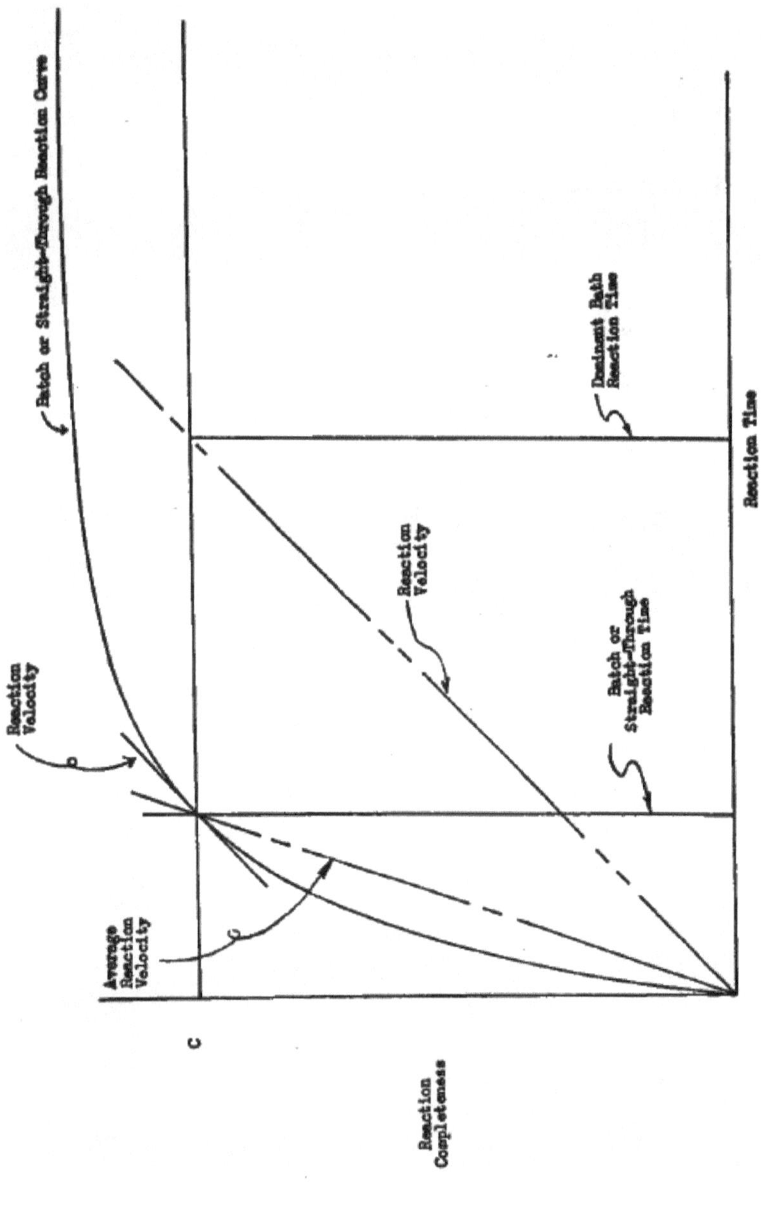

Kinetics of Straight-Through and Dominant-Bath Reactions

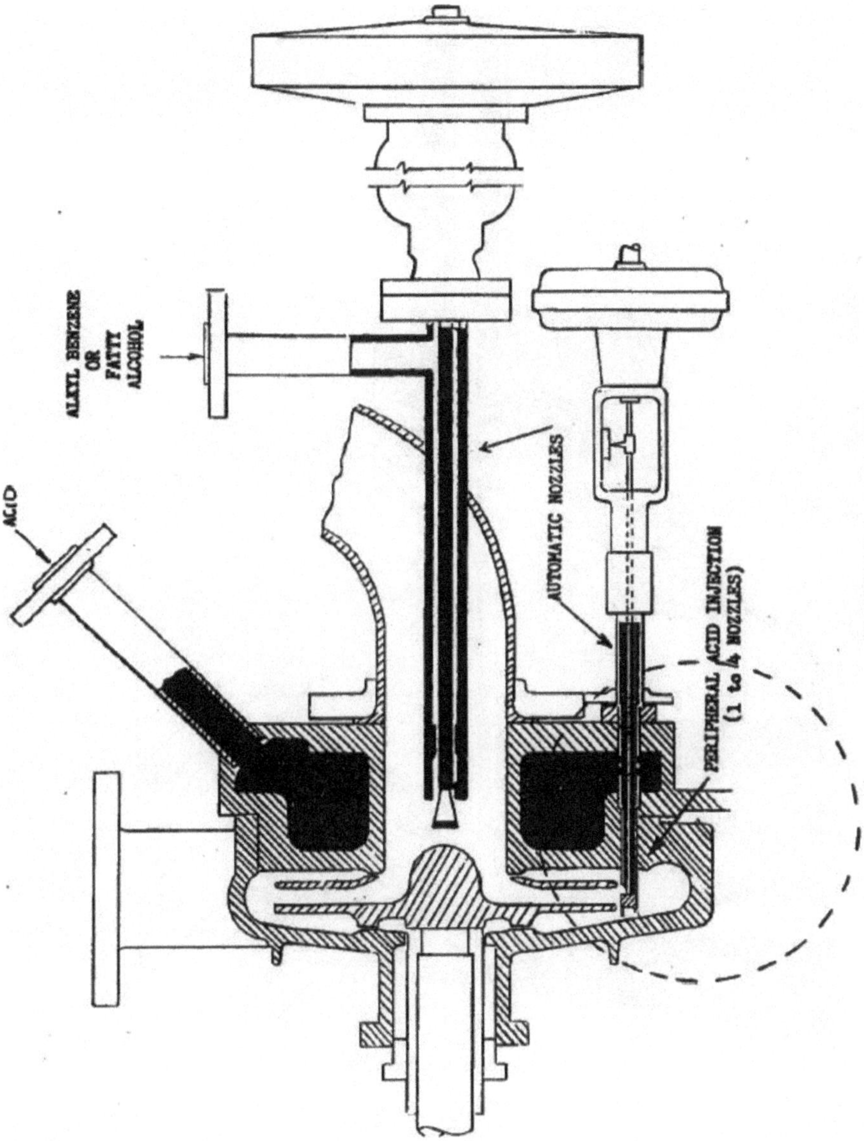

Peripheral Acid Injection

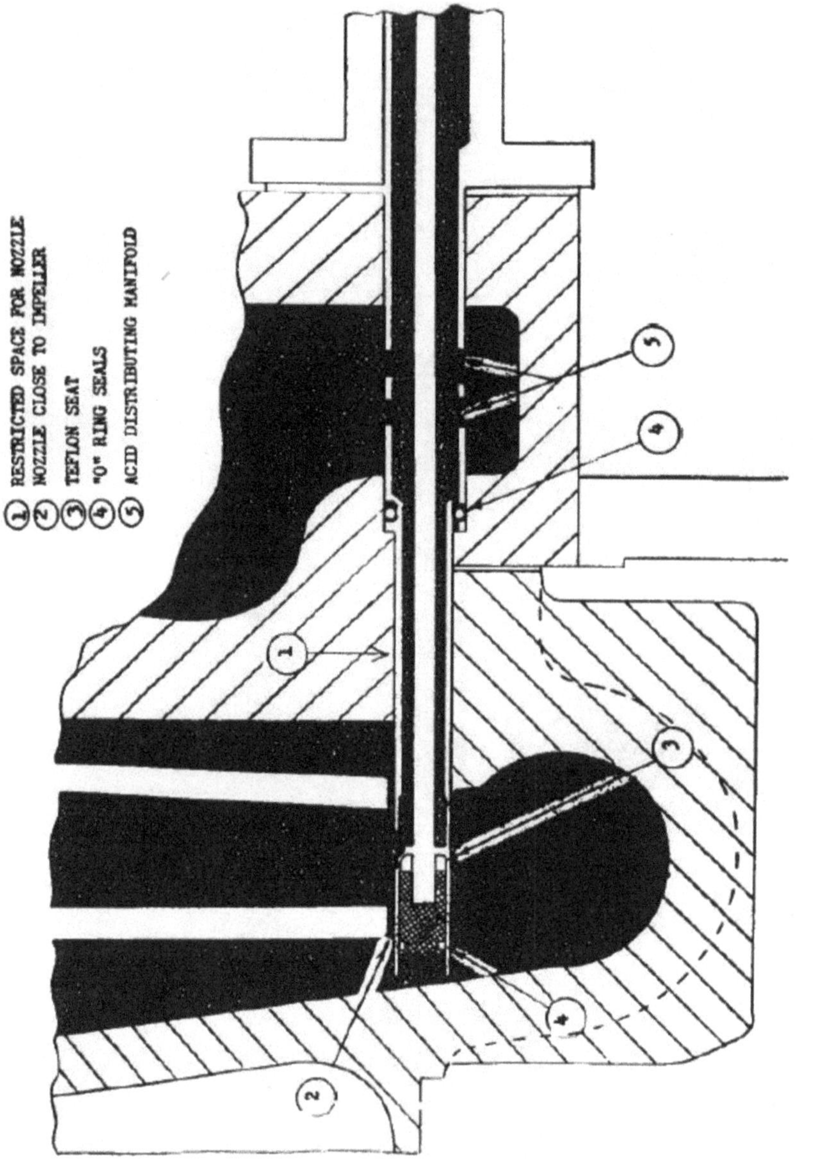

Peripheral Nozzle Critical Features

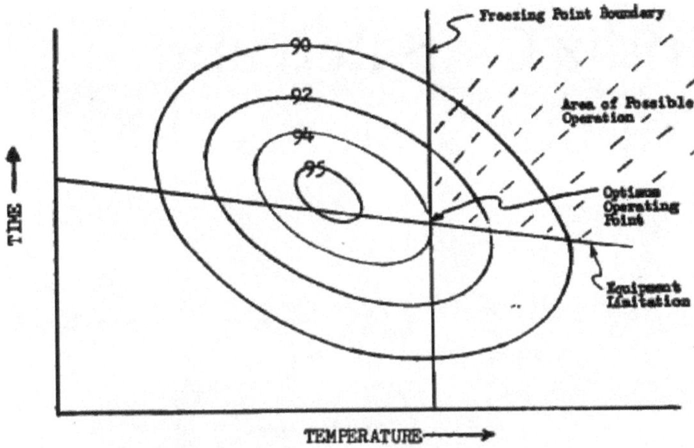

Contours of Sulfation Completeness in a Dominant Bath for Varying Time and Temperature at Constant Acid Ratio

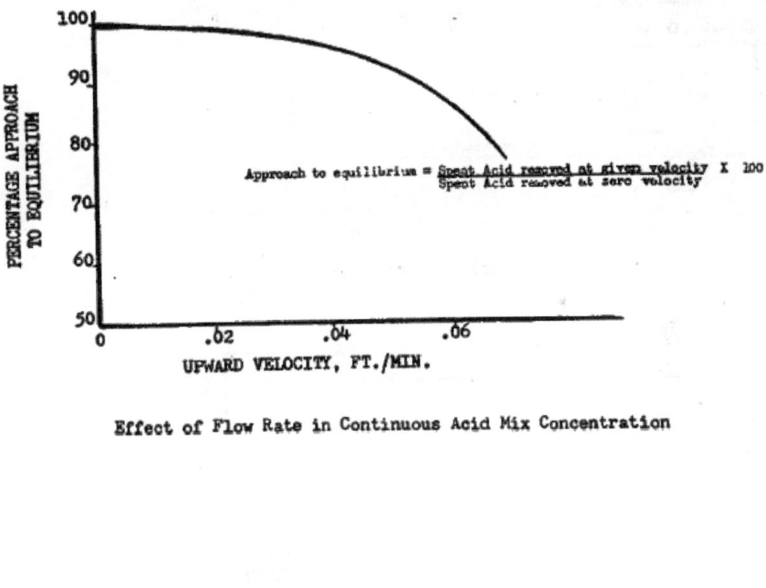

Effect of Flow Rate in Continuous Acid Mix Concentration

Appendix 6 Batch and Continuous Systems Flow Diagrams

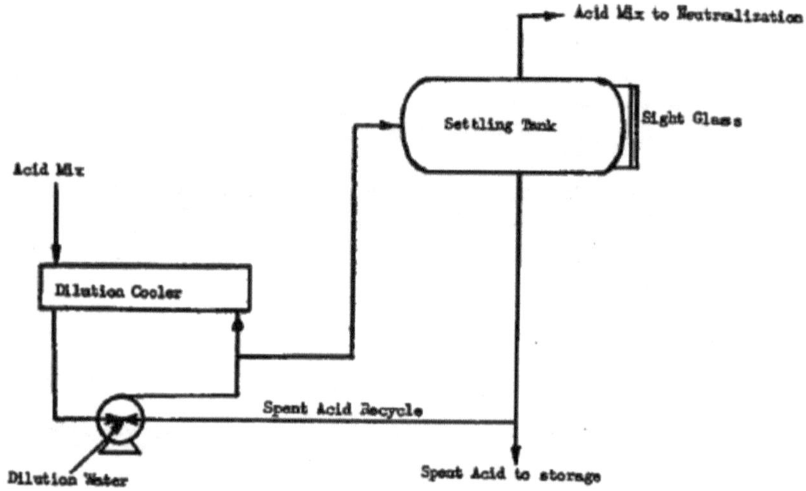

Continuous Settling System

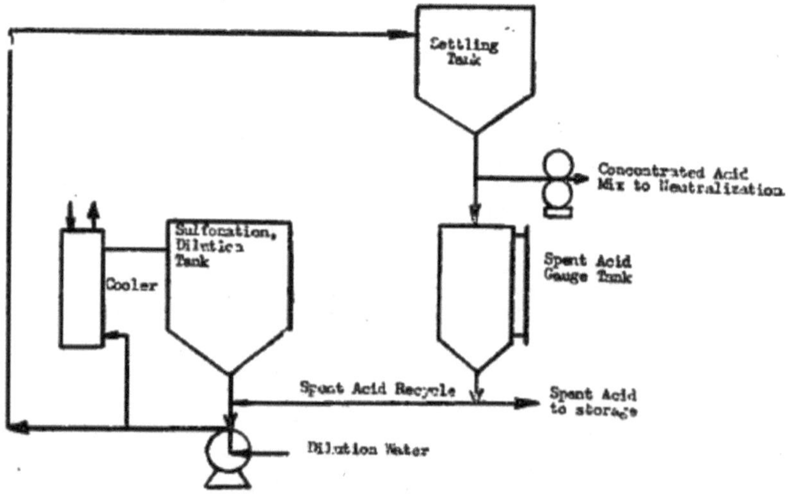

Batch Settling System

Appendix 7 Chemistry of Neutralization

<u>Chemistry of Neutralization</u>

The neutralization of acid mix is relatively simple:

1. $RO\text{-}SO_3H + NaOH \longrightarrow RO\text{-}SO_3Na + H_2O$

2. 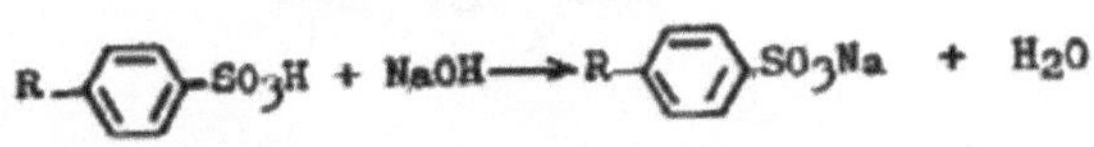

3. $H_2SO_4 + 2NaOH \longrightarrow Na_2SO_4 + 2H_2O$

These are the highly exothermic reactions of strong acids with strong base:

$$H^+ + OH^- \longrightarrow H_2O + 13.6 \text{ k-calories (dilute solution)}$$

NEUTRALIZATION

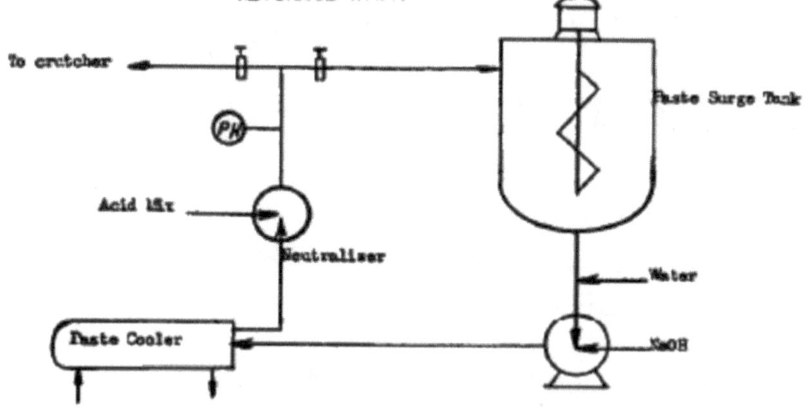

Continuous Neutralization, Short Time Lag
(Shown with recirculation pump used for paste transfer)

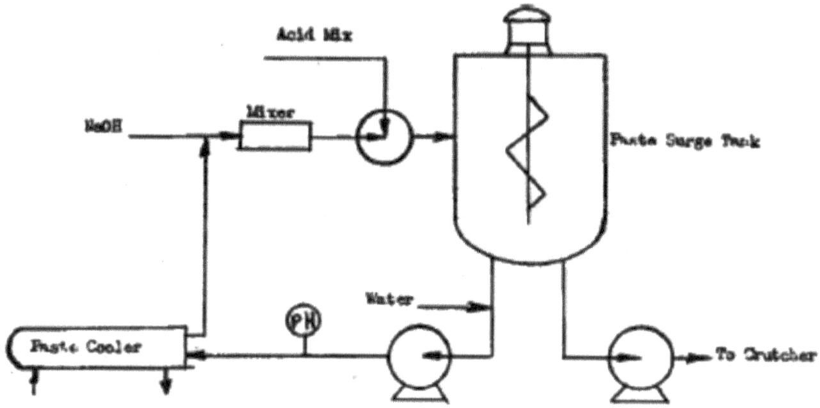

Continuous Neutralization, Large Time Lag
(Shown with paste transfer pump)

Appendix 8 Three Components Phase Diagram

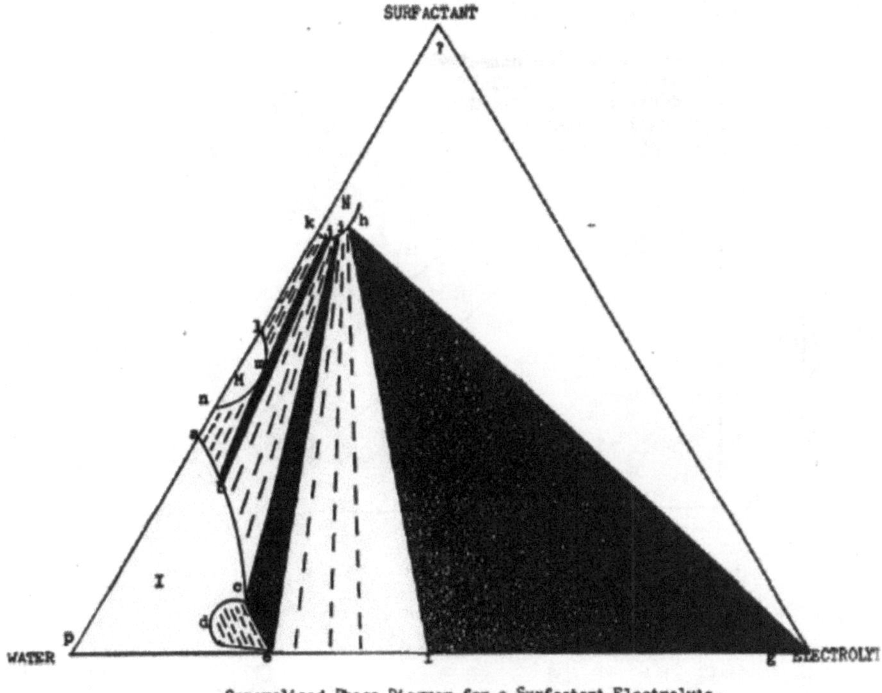

Generalized Phase Diagram for a Surfactant-Electrolyte-Water System

Appendix 9 Pressure Nozzles Spray

Reynolds number, Re = $\frac{dv\rho}{\mu}$, is plotted versus a dimensionless group Z whose importance can be predicted by dimensional analysis.

$$Z = \frac{\mu}{\sqrt{\sigma \rho d}}$$ where:

d = nozzle orifice diameter
v = velocity of the fluid
ρ = density of the fluid
μ = absolute viscosity
σ = surface tension

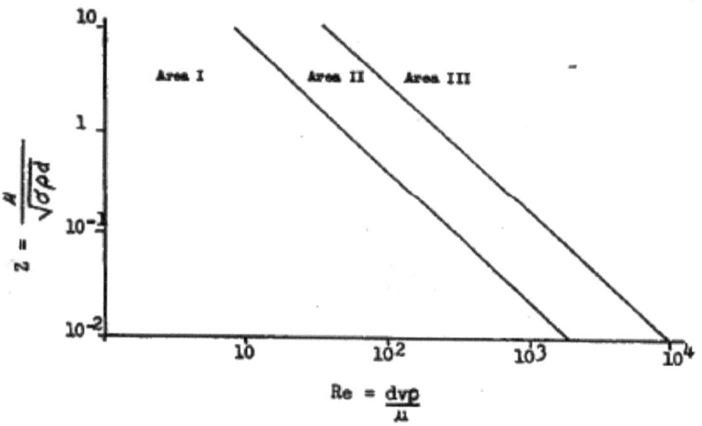

Spray Characteristics of Pressure Nozzles

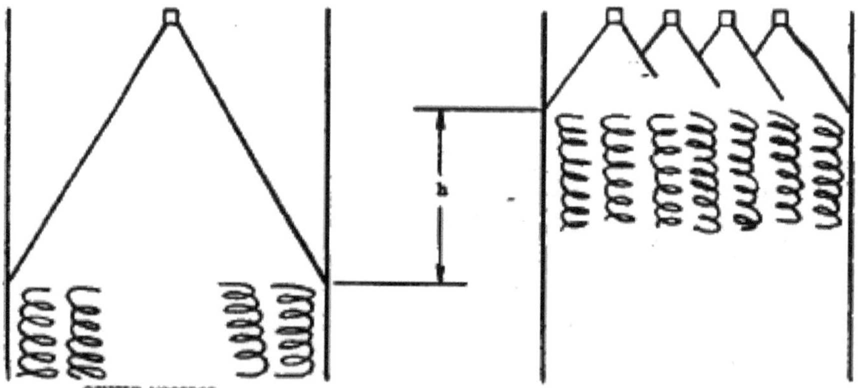

h = loss in effective tower height

Spray Pattern of Single vs Multiple Nozzles

Appendix 10 Spray-Drying Flow

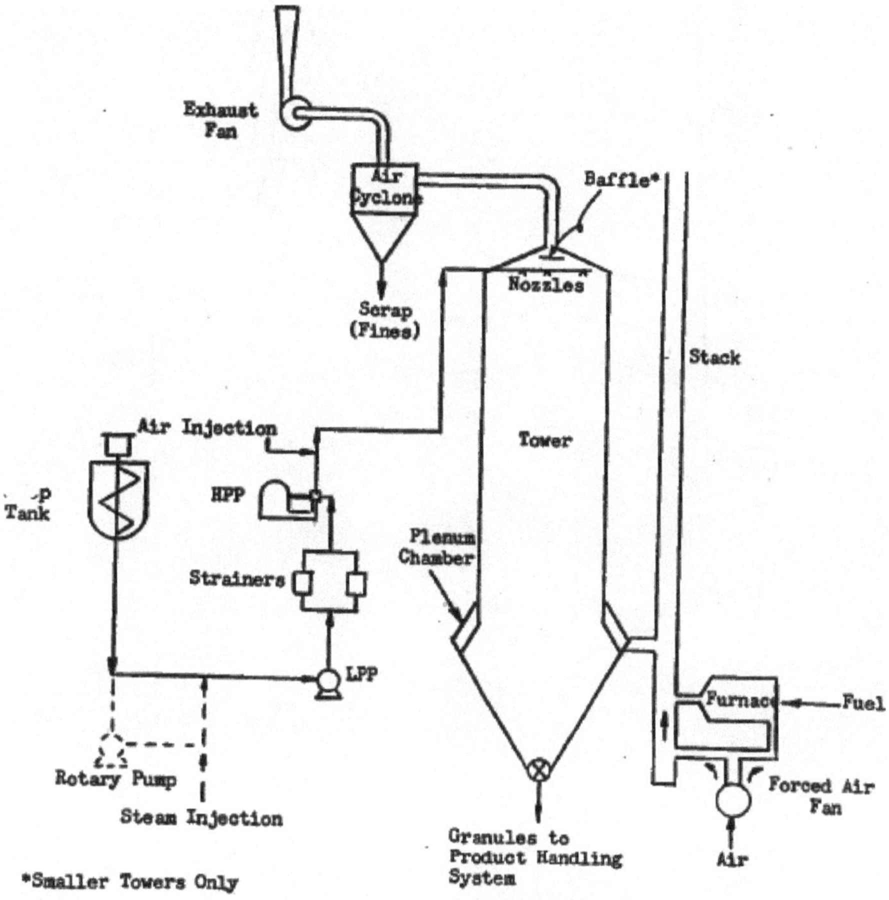

Flowchart of Counter-Current Spray Drying System

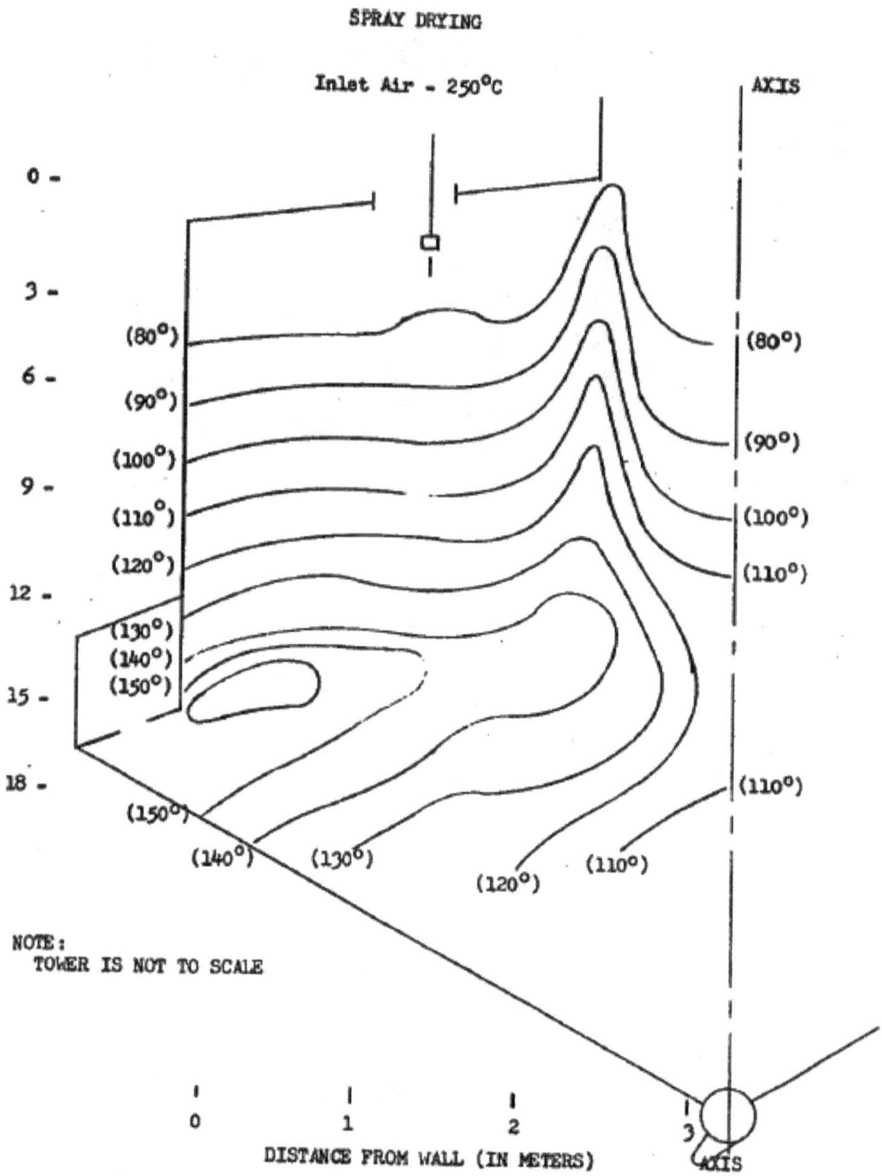

Typical Temperature Distribution in a Counter-Current Tower

SPRAY DRYING

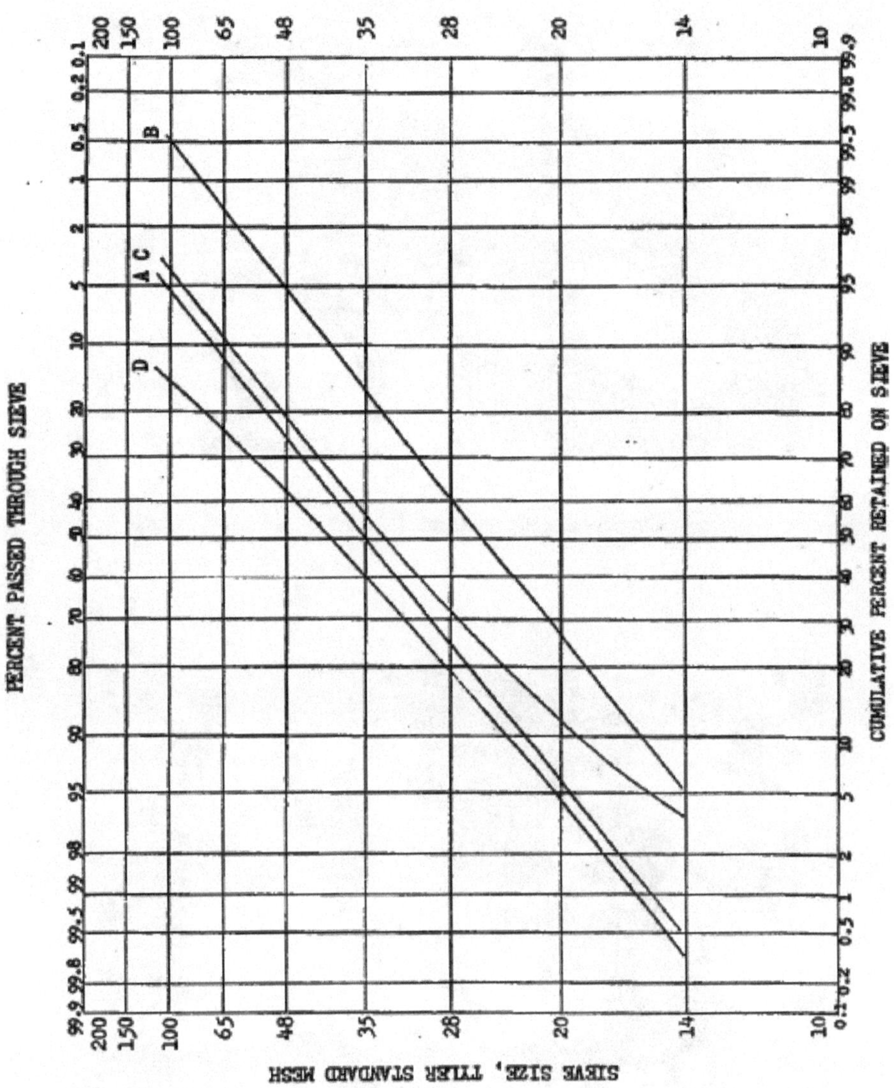

Typical Particle Size Distributions

(a) Countercurrent Operation (X40)

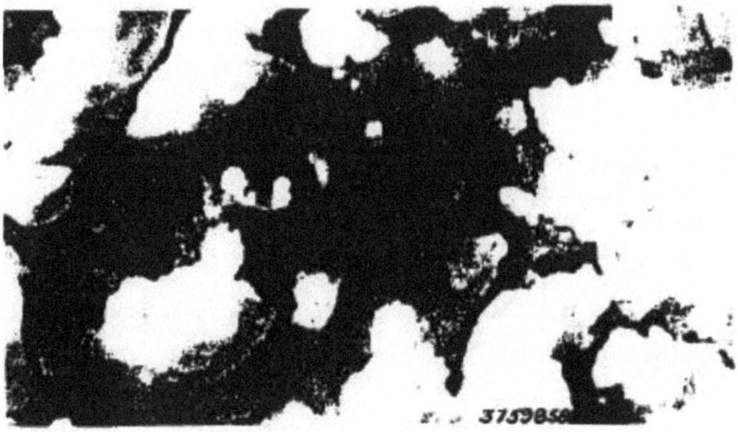

(b) Concurrent Operation (X40)

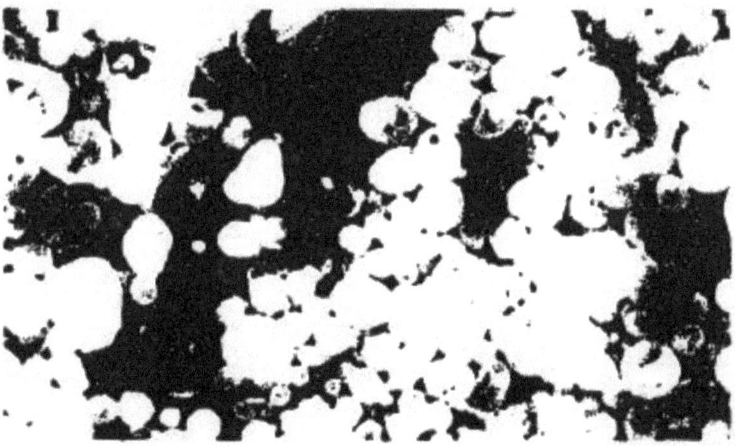

Typical Spray Dried Granules (35% TSD--35% STP)

Appendix 11 Detergent Product Handling

PRODUCT HANDLING

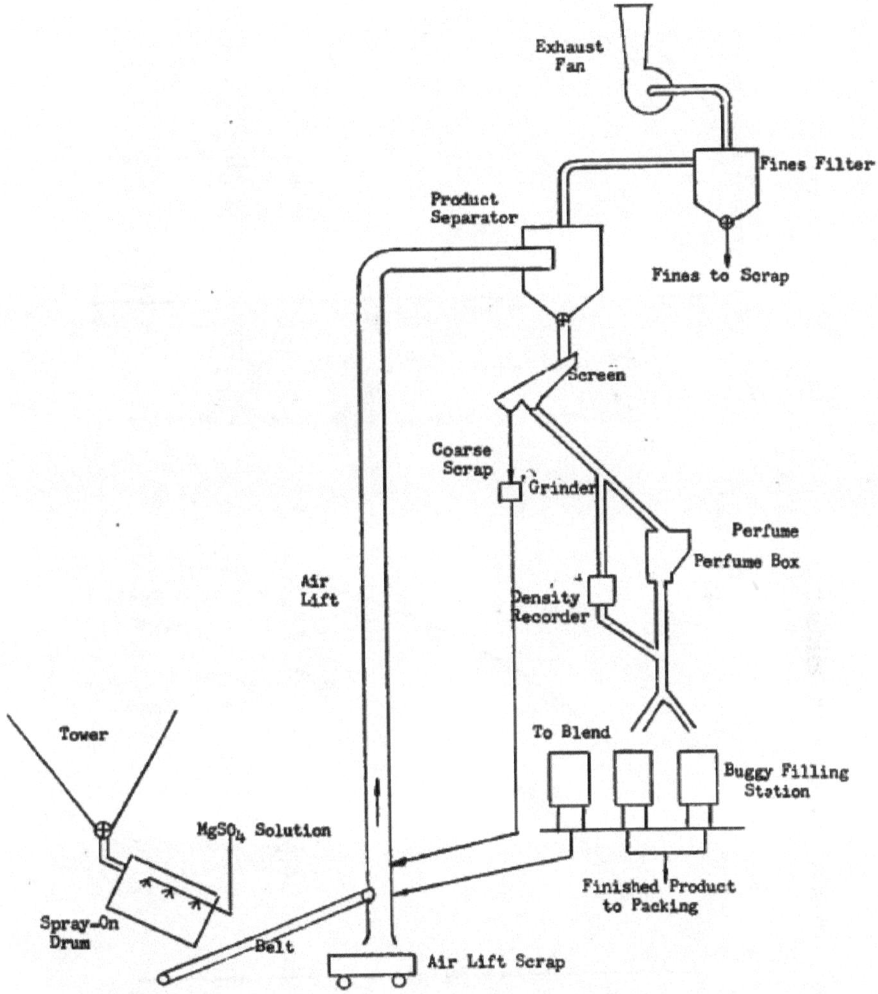

Diagrammatic Flow Chart of Product Handling System

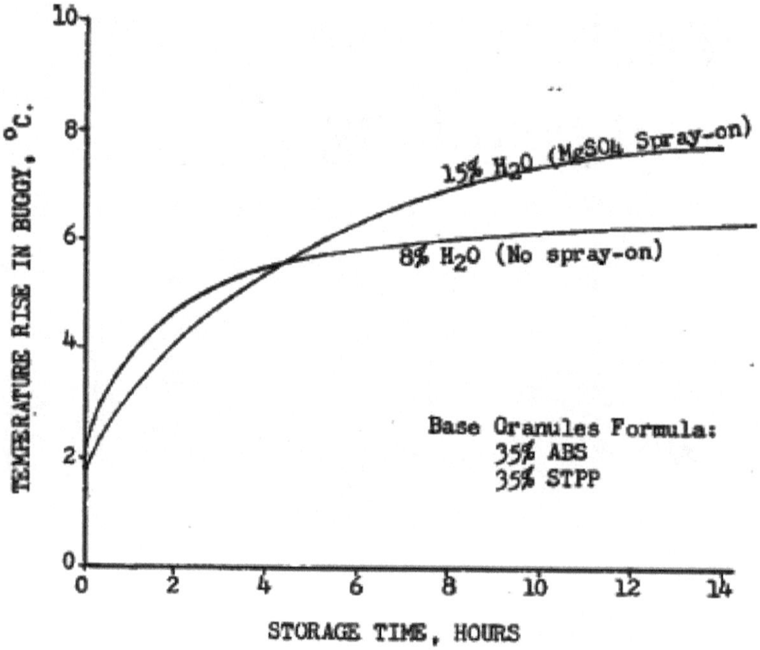

Temperature Rise of Granules in Buggies

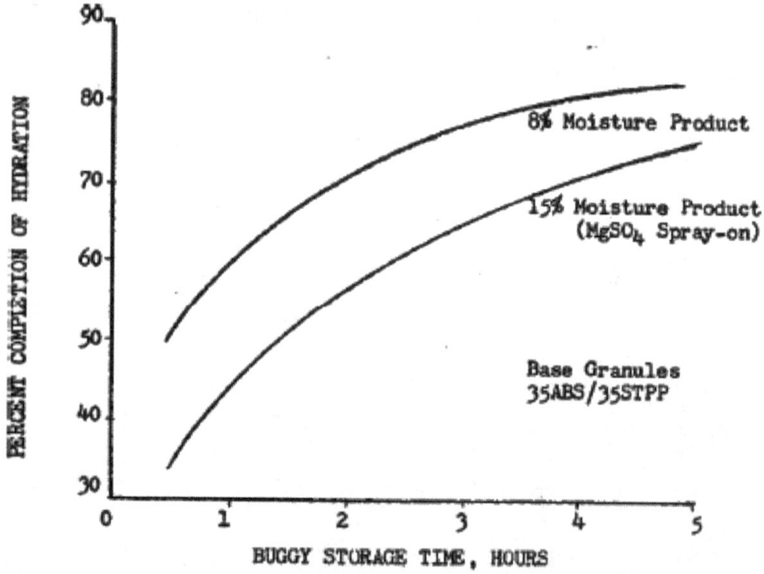

Rate of Hydration of Granules in Buggies

Appendix 12 Detergent Packing

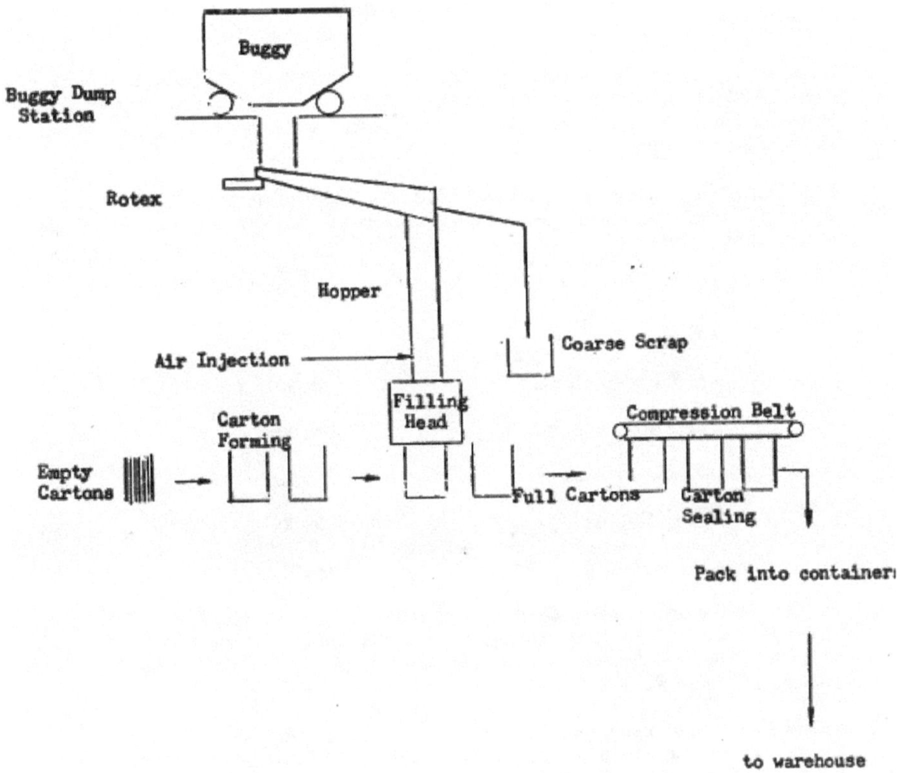

Flowchart of Carton Packing Operation

Appendix 13 Materials Balances

MATERIAL BALANCES

Rows A, B, C, D, E, F show the formation of finished product constituents ABS, free AB, STS, free toluene, Na_2SO_4, and Na_2CO_3 respectively Rows K, L, M, N, show the ingredients added during crutching. Rows R, S show the $MgSO_4$ spray on.

2. The finished product composition for the following constituents is fixed by performance or processing reasons. In the example chosen, these values are:

ABS	35.00
Toluene sulfonate	3.59
Tripolyphosphate	30.00
Silicate solids	6.00
CMC (100% Basis)	.225
$MgSO_4$ Solids	2.00
Moisture	15.00

3. To determine CMC usage, we must determine the purity of the CMC that will be used. This information is on the RMS. Assume that CMC purity is 66%. The CMC usage is .225/.66 = 0.34 parts. Enter this figure in column 15, row N.

4. Enter the other figures given in Step 2 in column 15. Include Miscellaneous gain in spray drying which for Carbonate Neutralization is 0.20 (Synthetics Factors).

5. Working from right to left, Line A is filled in by calculating the parts of AB required to form the indicated amount of ABS:

$$\text{AB required for ABS} = \text{ABS in fin. product} \times \frac{\text{mol. wt AB}}{\text{mol. wt ABS}}$$

$$= 35.00 \times \frac{249}{351} = 24.83$$

$$SO_3 \text{ required for ABS} = \text{ABS} \times \frac{\text{mol. wt. } SO_3}{\text{mol. wt. ABS}}$$

$$= 35.00 \times \frac{80}{351} = 7.98$$

$$\text{AB sulfonic acid} = \text{AB} + SO_3$$

$$= 24.83 + 7.98 = 32.81$$

	PASTE MAKING *	1 AB	2 TOL	3 30% Oleum SO_3	4 Water Added	5 Spent Acid Removed	6 Acid Mix	7 Spent Acid Returned	8 Na_2CO_3 Na_2O	9 Na_2CO_3 CO_2	10 Paste Loss CO_2	11 Na_2O	12 Water Formed	13 Paste Total	14 Tower Loss	15 Finished Product
A	ABS	24.83		7.98			32.81		3.09				-0.90	35.00		35.00
B	AB	0.50					0.50							0.50	-0.13	0.37
C	STS		1.70	1.48			3.18		0.57				-0.16	3.59		3.59
D	Toluene		0.30				0.30							0.30	-0.30	–
E	Sodium Sulfate			11.30		-7.50	3.80	0.14	2.93			0.13		7.00		7.00
F	Sodium Carbonate									4.67	-4.46	0.29		0.50		0.50
G	Water Free Total	25.33	2.00	20.76		-7.50	40.59	0.14	11.26		-4.46	0.42	-1.06	46.89		
H	Water			3.05	3.39	-4.28	2.16	0.08	44.81		-1.22		+1.06	46.89		
J	Totals	25.33	2.00	23.81	3.39	-11.78	42.75	0.22	56.07		-5.68	0.42		93.78 (50% H_2O)		

	CRUTCHING	16 TOTAL	17 SOLIDS	18 WATER		15
K	Paste	93.78	46.89	46.89		As above
L	STPP	30.00	30.00			30.00
M	Silicate	13.33	6.00	7.33		6.00
N	CMC	0.34	0.34			0.34
					Water	9.22
					Misc. Gain	0.20
O		137.45	83.23	54.22		
P				(39.5%)		
	SPRAY-ON					
R	$MgSO_4$	2.00	2.00			2.00
S	Water	7.78		7.78		
		9.78	2.00	7.78	(2% moisture lost in airlift)	5.78
						100.00

BASIS: Synthetic Factors #9

*AB molecular, weight = 249
AB completeness = 98%, at 0.94 acid ratio
Toluene completeness = 85%, using series sulfonation
All of ABS and STS neutralized by carbonate.
96% of sulfate formed from carbonate.

Appendix 14 Surfactants Processes

Sulfonation and Sulfation

Sulfonation is a reaction in which an organic material reacts with a source of SO_3 to form a sulfonic acid. When neutralized, this gives a sulfonate. The reaction takes place at an active hydrogen and the product has a direct C-S linkage (R-C-SO_3).

Sulfation is also a reaction of a material with a source of SO_3 but in this case an alkyl sulfuric acid is formed and the neutralized product is a sulfate ester. On sulfation, the reaction takes place at a hydroxyl group so that the product has a C-O-S linkage (R3-C-O-SO_3H).

The chemistry of sulfonation and of sulfation are different and are, therefore, discussed separately below. In addition, in making granular synthetic detergents, the materials which are most commonly sulfonated and sulfated are alkyl benzenes (AB), toluene, and fatty alcohols, respectively, so that the reactions of only these materials will be considered. The principal variables in sulfonation are:

(a) Temperature of reaction
(b) Time of reaction
(c) Acid strength
(d) Acid ratio
(e) Effectiveness of mixing

With suitable combinations of these variables, it is possible to fix the following most important reaction characteristics:

(a) Completeness of reaction
(b) Color and odor of the final product
(c) Ratio of sulfonic (or alkyl sulfuric) acid to H_2SO_4 in the acid mix

ACID STRENGTH

In this discussion, the term acid will refer only to H_2SO_4, oleum or SO_3. These are the most often used sulfonating agents for synthetic detergents. Chloro-sulfonic acid ($ClSO_3H$) has been used primarily in the preparation of actives for liquid detergents. Thus, the term "acid strength" refers to the SO_3 content of the sulfonating agent.

Acid ratio is a convenient means for expressing the acid usage. It is defined as the weight ratio of acid, as used, to AB or alcohol. From a theoretical standpoint, the term is meaningless unless the acid strength and the molecular weight of the AB or alcohol are known.

Acid mix refers to the complex mixture obtained from sulfonation. It consists of reaction products, unreacted AB or alcohol, excess acid, side reaction products, etc. Neutralized acid mix is called paste. Completeness is defined as the weight ratio of AB (or fatty alcohol) which has reacted to form AB sulfonic acid (or alkyl sulfuric acid) to the total AB (or alcohol) fed to the reaction mix. It is determined by analyzing neutralized paste for its sulfonated and sulfonated contents. Completeness cannot be determined by analyzing granules since part of the sulfonated material is evaporated during spray drying.

SPENT ACID

Normally an excess of acid over the theoretical requirement of SO_3 is used. In the acid mix, this excess acid is called spent acid and its strength is calculated as percent H_2SO_4. It is made up of the SO_3 not consumed in the reaction plus all the water which was introduced with the sulfonating agent. Spent acid is important because, when neutralized, it forms Na_2SO_4, which remains in the product. Spent acid strength is a function of acid strength, acid ratio, completeness, and molecular weight of the AB or fatty alcohol.

CHEMISTRY OF SULFONATION

SULFONATION REACTION

The desired sulfonation reaction is:

Alkyl benzene + SO_3 = Alkyl benzene sulfonic acid

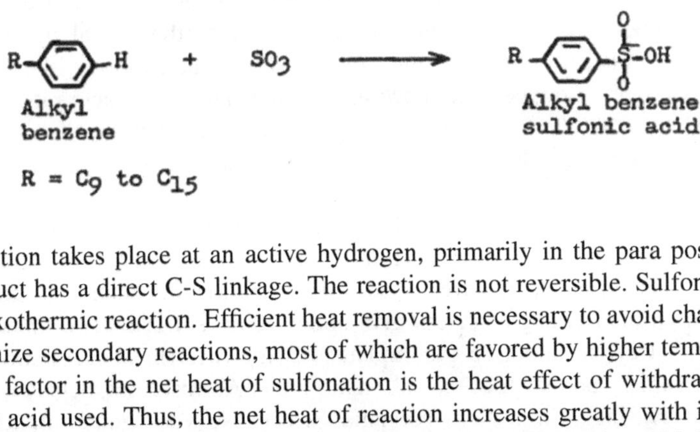

The reaction takes place at an active hydrogen, primarily in the para position and the product has a direct C-S linkage. The reaction is not reversible. Sulfonation is a highly exothermic reaction. Efficient heat removal is necessary to avoid charring and to minimize secondary reactions, most of which are favored by higher temperatures. A major factor in the net heat of sulfonation is the heat effect of withdrawing SO_3 from the acid used. Thus, the net heat of reaction increases greatly with increasing strength. For a given acid strength, the small rise of heat of reaction with increasing spent acid strength reflects increased completeness and the effects of side reactions. With 25% oleum and normal acid ratios, typical heats of reaction are:

97 cals/g AB in sulfonation
80 cals/g tallow fatty alcohol in sulfation
97 cals/g coconut fatty alcohol in sulfation

COMPLETENESS OF SULFONATION

Since sulfonation is an irreversible reaction, given sufficient time and sulfonating agent the completeness obtained is limited only by the purity of the AB (commercial ABs contain about 1% of un-sulfonable material). Side reactions except under extreme conditions do not measurably affect completeness. Only rate of reaction and not the maximum degree of completeness obtainable is a function of acid strength.

ACID RATIO

Completeness of alkyl benzene sulfonation, in practice, is whereby an excess of acid over the stoichiometric requirement is always used for sulfonation so that, neglecting the small effect of side reactions, the actual completeness obtained will depend on the rate of sulfonation. This, in turn, is a function of reaction temperature. Increasing the sulfonation temperature increases the rate of reaction. This effect is small, however, above about 45 °C except when the acid ratios and acid strengths used are such that the spent acid strength drops to about 98% H_2SO_4 or less.

During the initial stages of sulfonation (for example, until about 80–85% of the available AB in a batch system has reacted), rate of sulfonation is a function of the unreacted AB and SO_3 concentrations. Except when sulfonating with pure SO_3, a considerable molar excess of SO_3 is used. The unreacted SO_3 concentration, therefore, does not change greatly during sulfonation and the reaction rate depends, primarily, on the concentration of unreacted AB.

There are two instances where SO_3 concentration is important. First, there is a critical spent acid concentration π value below which the reaction does not proceed at any appreciable rate. This is a limiting concentration and the rate is exceedingly slow at this concentration. The value appears to have no theoretical significance in itself especially since it varies with acid strength. Normal acid ratios will maintain the spent acid strength a few percent above the π value. The actual acid ratios required are calculated to obtain 98% completeness, with normal reaction times, for various strength sulfonating agents. The corresponding molar ratios of SO_3: AB and the resulting spent acid strengths are also calculated.

The mechanism of sulfonation at low acid strengths (less than about 90% sulfuric acid) is believed to be different from the mechanism with oleum (about 100%, and higher). Experience, however, shows that reaction rate increases sharply in going from an acid strength of less than about 100% sulfuric acid, to one above this strength. Beyond about 85% completeness, the reactant concentrations begin to decrease rapidly and so does the sulfonation rate. At this point, degree of mixing becomes increasingly important, presumably because very good mixing of the viscous acid mix is necessary activity of sulfuric acid bringing the two reactants, AB and SO_3, together for reaction. Thus, in normal operation, in the completeness range of interest (97–98%), mixing, rather than reactant concentrations, controls the overall reaction rate.

The importance of effective mixing is illustrated by data. These data are for a special continuous sulfonation with no external reaction coil and are relative only. Increasing speed of mixing may be taken as better mixing; the effect on completeness is evident. Side reactions, such as changes in color and odor of the final product, are observed. Several secondary reactions also occur during sulfonation:

(a) Di-sulfonation probably occurs slowly by reaction between SO_3 and the alkyl benzene sulfonic acid formed in the primary reaction. Although this reaction is believed to take place, it has not been confirmed by paste analyses. The amount of di-sulfonate formed is certainly small and is probably unimportant.
(b) Oxidation and dehydration reactions take place largely under conditions of high SO concentration and high temperature. The removal of water by SO is the driving force in this reaction. Most other side reactions are also favored by high SO concentrations; consequently, it is preferable to add acid to AB rather than vice versa.
(c) Rearrangement of the alkyl chain probably occurs to some extent during sulfonation, but this has not been studied in detail. Many of the factors which increase the rate of sulfonation also increase the rate of the side reactions. As previously noted, side reactions reduce the completeness obtainable. In addition, however, side reaction products may cause unsatisfactory color and odor even when present in such small amounts that their effect on completeness is negligible. As with sulfonation, the rate of the side reactions decreases rapidly as the acid strength drops below 100%, sulfuric acid presumably again, because the activity of the sulfuric acid drops at this point. When 20–30% oleum is used as a sulfonating agent, the maximum allowable sulfonation temperature for good color is approximately 35 °C, whereas with 100% sulfuric, the acid mix is far less sensitive to color degradation, and temperatures up to about 60–70 °C are used.

Although the mechanism of color formation is not fully understood, it probably results from side reactions which occur at the surfaces of the acid droplets formed upon injection of the acid into AB or acid mix. Since the reactions are exothermic, local overheating at the acid droplet surfaces accelerate the charring and other undesirable side reactions. Good mixing to dissipate the high local concentrations of SO and to prevent high local temperatures is of major importance in minimizing color formation. Effectiveness of mixing determines, to a large extent, the maximum acid strength, which can be used without obtaining poor colors. The better the mixing, the higher is the acceptable acid strength.

Under normal conditions of acid ratio, acid strength, temperature, mixing side reactions are generally not a problem. Difficulty with color can occur, however, if any of these factors is out of control. This happens most often with acid ratio or temperature during startups or shutdowns.

The desired sulfation reaction is:

$$ROH + \text{Fatty Alcohol} = SO_2R\text{-}0\text{-}S\text{-}OH$$

The reaction takes place at a hydroxyl group so that the product has a C-OS linkage. In the presence of water, this reaction is reversible by hydrolysis. Sulfation with sulfuric acid or with oleum is an equilibrium reaction, the maximum possible completeness being limited even without consideration of side reactions. However, because of the reactivity of fatty alcohol, the side reactions are more numerous, more serious, and more rapid than in sulfonation. These are, again, dehydration and oxidation reactions. Some typical side reactions are:

$$RCH,CHOH + SO_5$$

$$2RCH,CHOH + SO_5$$

$$2RCH,CHOH + 4SO_4$$

$$RCH,CHOH + SO_4$$

$$RCH = CH + H_2SO_4$$

$$RCH,CH,OCH,CH,R + H_2SO_4$$

$$C\text{-}O\text{-}C,CH_3R + 2SO_2 + 2SO$$

The un-sulfated portion of an alkyl sulfate paste, obtained by extraction with petroleum ether, was analyzed. The incompleteness is primarily what may be called over-sulfation rather than under-sulfation. The term "un-sulfated" is, however, commonly applied to this portion of a paste. Although the sulfation reaction is very rapid, the alkyl-sulfuric acid is not stable. Both the alcohol and the alkyl-sulfuric acid undergo degradation reactions with oleum. The net effect is that the completeness, as measured by the amount of alkyl-sulfuric acid obtained, has a maximum at a rather short reaction time.

Optimum conditions for sulfation are based upon obtaining high primary reaction completeness and terminating the reaction to minimize the secondary reactions. The sulfation reaction is terminated by neutralizing the acid mix since alkyl sulfate is relatively stable at a pH above 7. Since alkyl sulfuric acid hydrolyzes in the presence of water, neutralization is usually accomplished by adding acid mix to the neutralizing agent (NaOH or Na_2CO_3). The heat of neutralization must be removed to avoid decomposition of the alkyl-sulfuric acid.

PROCESSING

Sulfonation is carried out in both batch and continuous systems. Sulfation is generally a continuous process, although batch systems are used when SO or chloro-sulfonic acid is the sulfating agent.

BATCH SULFONATION

A typical batch sulfonation unit consists of a vessel equipped with a means of mixing the AB and acid and of removing the heat of reaction. Tanks with cooling coils or jacket, for example, and an agitator may be used. A more effective system, however, consists of a vessel with an external heat exchanger through which the acid mix is circulated by means of a recirculation pump.

All the AB is charged to the tank and recirculated through the cooler as acid is slowly added. The acid is conveniently added at the pump, where most of the mixing takes place. An agitator is generally also provided in the vessel but serves principally to maintain homogeneity within the vessel. Addition of acid is slow (30–60 minutes per batch) to ensure good mixing of the acid as it is added and to minimize the load on the cooling equipment. After all the acid has been added, an additional aging period (30–60 minutes) is required to allow the sulfonation to go essentially to completion.

In such batch systems, mixing tends to be relatively poor. Completeness is often lower than is obtainable by other processes and the finished product colors are somewhat poorer. The batch system has the advantages, however, of requiring the cheapest and simplest equipment and of providing the greatest ease of operation with a minimum of process control problems. There are no metering difficulties and few equipment problems. Actual rate of acid addition is not important provided only that the maximum permissible rate is not exceeded. Total amounts of AB and acid added are readily controlled and so is the temperature.

SEMI-CONTINUOUS SULFONATION

The relatively poor paste color and completeness obtained in batch sulfonation generally arise from two factors:

(A) Poor Mixing

In a batch system, the acid passes through the circulation pump initially and then must reside in the tank for more than a minute before it reaches the mixing action of the pump again. Let most of the mixing of ingredients in this system occur in the pump; very little is due to the tank agitator.

(B) Reactant Concentrations

In the early stages of a batch sulfonation cycle, acid is added to pure AB. Much of the sulfonation, therefore, occurs by the direct action of high strength acid on the AB. This condition promotes side reactions and poor color. Not until later in the course of sulfonation are ABS and spent acid present in large enough quantities to disperse and dilute the AB and acid.

A so-called "semi-continuous" sulfonation provides improved reaction conditions over those usually obtained in batch systems. The only additional equipment needed is flow meters for AB and acid and a by-pass line from the discharge of the cooler to the suction of the recirculation pump. Both AB and acid are metered manually into the sulfonator pump. The acid mix formed is recirculated through the dominant bath loop consisting of the pump, a cooler, and the tank by-pass line. Initially, the sulfonation tank is empty but, during the course of the sulfonation run, it gradually fills with the acid mix from the dominant bath. After each batch, the lines are normally left full of acid mix. To drive out any possible air pockets, which would cause pump cavitation and lack of proper mixing, some AB (or acid mix from the preceding batch) is run into the system at the start of each batch until the external system is completely full of liquid. To minimize the side reactions which result from high SO concentrations, the rate of acid addition is set to give a slightly lower acid ratio than is finally desired. This makes control of the addition rates less critical by reducing the danger of instantaneous over-concentration of 50%.

As a result of this precaution, the AB addition is completed first and the acid addition is then completed as in normal batch sulfonation. That is, the acid mix is recirculated through the cooler and directly back to the tank during addition of the final portion of acid. A 30-minute aging period for the acid mix then follows.

In the semi-continuous procedure, most of the reaction occurs in the dominant bath. Here, mixing is greatly improved and, since both AB and acid are quickly dispersed in the recirculating stream of acid mix, sulfonation occurs, to a large extent, by the action of spent acid on AB. These improvements increase completeness by as much as 20% over the batch system and noticeably improve color.

CONTINUOUS SULFONATION: (DOMINANT BATH)

Heat removal and mixing are the major problems involved in making the sulfonation process continuous. Both difficulties are readily solved by a dominant bath process. In a dominant bath process, the reactants are injected into a large mass of reaction products, the relative magnitude of the latter being sufficient to be predominant. The heat of reaction is thus dissipated into the large mass of material without excessive temperature increase, and the larger volume facilitates heat removal and mixing.

Various types of dominant bath systems are used. A common dominant bath system is the continuous, stirred-tank reactor. Any batch unit can be converted to continuous use by providing accurate metering equipment for the AB and acid and a suitable overflow for the acid mix. The Ballestra sulfonation systems are of this type. The rest of this discussion, however, concerns the designed dominant bath which is both simpler and more efficient.

In this design, the reaction products (acid mix) are continuously pumped in a dominant bath loop consisting essentially of a heat exchanger, piping, and a pump. AB and acid are metered to the pump as in the semi-continuous system and acid mix is continuously removed. Here again, most of the mixing occurs within the pump. A back-pressure control valve maintains a positive pressure on the system at all times to prevent air entrainment and pump cavitation.

In an "ideal" dominant bath all parts of the bath are at the same composition, the composition of the effluent stream. The mean reaction time is equal to the system volume divided by the rate of volume throughput. In an actual dominant bath loop, the "recirculation ratio" (the flow rate of recirculated material divided by the flow of reactants) determines the degree of approach to ideality. For example, at a 5:1 ratio, one-sixth of the feed will be removed the first time past the take-off point, thus short circuiting the system and having a reaction time considerably less than the calculated mean. However, at a recirculation ratio greater than about 5:1, departure from ideality is negligible. If the chemical reaction rate is controlling, the rate is proportional to the concentration of unreacted materials. In a batch reaction, the reaction is fast initially since reactants are high in concentration. Then, as the reactants are consumed, the rate decreases. In a dominant bath, the reactants are always at the final concentration and hence the chemical reaction is relatively slower.

This fundamental feature of the dominant bath system, considering first the batch reaction curve, is the instantaneous reaction velocity as discussed above since it is dependent upon the concentration of the reactants. This instantaneous reaction velocity is the slope of the curve, that is, the change in completeness with time. The average reaction velocity would be expressed by the slope of the line connecting any point on the curve with the origin. However, in an ideal dominant bath all reactants are at the final concentration so that the reaction velocity would be expressed by the slope of the batch reaction curve at the particular completeness being considered. A line from the origin and parallel to the tangent of the batch reaction curve at that particular completeness would give the average reaction velocity or the effective reaction velocity in the dominant bath. Where this line reaches the level of completeness being considered, the reaction time may then be noted by referring to the abscissa. It is obvious that at any degree of completeness, a considerably greater reaction time is required in the dominant bath and that the higher the completeness being considered, the greater the difference in reaction time required for dominant bath as compared to batch reaction. The term "straight through reaction" has been used and is indicated to be the same as the batch reaction. This term is applied to a continuous process in which there is no end-to-end mixing and thus each increment of the mixture may be considered as a very small batch. While the problem of heat removal is readily handled by means of the dominant bath system, the reaction time requirements for high completeness in a dominant bath may be excessive. Thus, a combination of dominant bath and straight through sections is advantageous.

The required reaction time in the dominant bath (and hence the volume) can be reduced by operating at a lower completeness since the concentrations of reactants will then be higher. This is feasible if the desired final completeness is obtained by providing additional reaction time after the dominant bath. This is the function of a reaction coil. The coil minimizes end-to-end mixing, and thus we have the straight-through reaction which has, as stated above, the same reaction kinetics as a batch. The temperature rise in the coil is generally small.

Of importance is the division of reaction time between dominant bath and coil (straight through section). The abscissa is the total reaction time, dominant bath plus straight through. Actual plant systems give higher completeness due to the better mixing obtained on the larger units.

Toluene is a short chain alkyl benzene, having one methyl group attached to the benzene ring. Since its molecular weight is lower than regular alkyl benzene, it will take up considerably more SO than will an equal weight of AB. Furthermore, toluene is more reactive than AB and, in a mixture of the two (co-sulfonation), is sulfonated preferentially, so that any deficiency in acid results in a poorer completeness for the AB than for the toluene.

Toluene, however, is readily sulfonated by 98–100% HSO. Consequently, when minimum oleum usage is desired (for either economic or formulation reasons), "series" sulfonation of AB and toluene can be utilized. The AB is sulfonated first using all the acid. The toluene is then added.

The toluene completeness may not be as high as with co-sulfonation, but since un-sulfonated toluene is almost entirely evaporated in the spray tower, it will not cause the stickiness and performance problems that occur with high un-sulfonated AB. Series sulfonation of toluene can be co-sulfonation of toluene carried out in either the batch, semi-continuous, or continuous units.

In the batch or semi-continuous units, toluene is added after the AB aging period. In the continuous system, a separate sulfonator is required.
The process involves using an AB dominant bath and a toluene dominant bath, and an in-line mixer may be used.

SERIES SULFONATION OF TOLUENE

The following data illustrates the advantages of series sulfonation using 25% oleum:
Sulfonation of Toluene, Co-Sulfonation, Series Sulfonation

Acid ratio	1.20
AB completeness	97% (approx.)
Toluene completeness	98%
Spent acid strength	1.05

Note that toluene completeness drops while AB completeness remains high. This shows that toluene will not be sulfonated by alkyl benzene sulfonic acid. Minimum acceptable acid ratios for sulfonation of toluene is limited by the requirement that toluene completeness not be too low. Below 80%, the residual un-sulfonated toluene contributes a noticeable odor to the finished product. The reaction time required for toluene sulfonation is less than that for AB, but quantitative data are not available as yet. Toluene co-sulfonation, when used, is carried out under the same reaction time and temperature conditions as for AB alone (3 minutes minimum in the external coil, 5 °C, maximum).

With series sulfonation, the acid available for toluene sulfonation is only 98–100% instead of being oleum as in co-sulfonation. Consequently, the rate of toluene sulfonation is lower. The minimum reaction time in the toluene external coil must, therefore, be increased (minutes minimum). Since the rates of the various sulfonation side reactions are similarly reduced, series sulfonation can be performed at higher temperatures (up to about 65 °C). The higher temperature increases the rate of toluene sulfonation but does not hurt product color or odor. The higher allowable temperature also permits substitution of an in-line mixer for the entire toluene dominant bath system, since toluene sulfonation raises the acid mix temperature only about 10 °C (from about 50 °C to about 60 °C). Toluene is more susceptible to side reactions than is AB. Ortho and meta toluene sulfonic acids, di-sulfonic acids, and di-tolu sulfones may be formed. These compounds can cause stickiness in the spray-dried granules. Their formation is prevented by keeping the spent acid strength below 99%.

The continuous dominant bath loop, although not ideal, is the only practical means of providing the efficient cooling and short reaction time required for sulfation of fatty alcohols with oleum. (Sulfation with SO_3 or chloro-sulfonic acid, however, are generally performed in a batch reactor.) Equipment limitations still impose restrictions upon minimum reaction times. Further, the high concentration of alkyl sulfuric acid and low concentrations of alcohol in the dominant bath favor the secondary reactions relative to the desired reaction.

Index

A

AB, 40
ABS, 40
ABS crutcher mix, 67
Acid ratio, 34, 46
Acid strength, 34, 46, 51
Active ABS paste, 66
Actives, 19
Alkylbenzene sulfonate, 11, 36
Anionics, 8
Anisotropic liquid phases, 65
AS, 69
Association Internationale de la Savonnerie et de la Detergence, 1

B

Bacteriostats, 30
Batch process, 54, 172
Batch sulfonation, 39
Biodegradability, 7
Bleaching systems, 2
Brighteners, 28
Builder, 10

C

Caustic strength, 57
Continuous sulfonation, 43
Continuous process, 54, 172
Crutcher Mix, 64
Crutching, 70
  atomizability, 70
  batch crutcher, 83
  continuous crutchers, 83
  mechanism of hydration, 72
  puffability, 71
  STPP, 72, 74
  STS, 76

D

Detergent intermediates capacity, 4
Detergent raw materials, 13
  sodium perborate, 13, 18, 27
  sodium silicate, 18
  sodium sulfate, 18
  sodium toluene sulfonate, 13
  Suds depressor, 3

E

Enzymes, 32

F

Fatty alcohol, 2, 7, 15
Finished product formulations, 53
Finished product quality, 140
  color, 140
  composition, 144
  density, 142
  finished products performance, 146
  grit, 143
  lumps, 141
  odor, 141
  particle size, 143
  pour grade, 143

G

Global detergent formulations, 3, 5

H

Hard-water detergent, 9
Heat of sulfonation, 21
Heavy-duty liquids, 2

L

Low sudsing, 11

M

Market, 147
  effect of moisture, 149
  humidity caking, 147
  product age, 154
  stacking, 148
  weather conditions, 152
Material balances, 156, 184
Mixing, 37, 47, 48
Multi-phase system, 66, 175

N

Neutralization, 56, 173
Nonionics, 8

O

Oxygen bleaching, 7

P

Packing, 134, 183
  buggy dumping, 134
  carton packing machines, 135
  coding, 138
  envelope and bag packing machines, 137
  hopper, 135
  premiums, 138
  screen, 135
  weight and outage control, 137
pH control, 59
phase separation, 63, 175
Phosphates, 8
Process control, 47
Process variables, 132
Product handling, 106, 181
  airlift, 108
  bin filling station, 112
  coarse scrap grinder, 111
  conveyor belt, 107
  density recorder, 111
  evaporative cooling in the airlift, 115
  inspection and cleanout programs, 117
  perfume box, 112
  phosphate hydration and temperature rise in granules, 113
  product separator, 109
  product sifter, 110
  reaction temperature, 49
  scrap handling, 116
  spray-on drum, 106

S

Soap, 9
Sodium bicarbonate, 8
Sodium carbonate (soda ash), 61
Sodium carboxymethyl cellulose, 11
Sodium nitrilotriacetate (NTA), 4, 6
Sodium tripolyphosphate, 10, 17, 24
Special processing, 119
  agglomeration, 125
  bleaching, 120
  caking, 120
  color-solution spray-on, 126
  metering, 122
  mixing and segregation, 122
  perfume box, 130
  perfume spray-on, 128
  sampling for process control, 123
  scrap handling, 123
  speckling, 125
Spray drying, 90, 91, 177
  agglomeration, 105
  density, 100
  fragments of granules, 103
  particle size, 101
  specks, 102
  spray-drying process control, 99
  spray-drying towers, 98
  steam injection system, 96
  sticky granules, 103, 104
Spray nozzles, 85, 89, 176
  pressure nozzles, 86
  spinning disk, 89
  two-fluid, 86
STS, 68, 76
Suds builders and suds depressors, 26
Sulfation, 34, 44, 168, 186
Sulfonation, 12, 34, 166, 168, 186
Surfactants, 7, 14, 15, 165, 186
  alkyl benzene, 16
  cationic, 14
  ionic, 14
  nonionics, 15
  oleum, 16
Syndets ingredients, 164
Synthetic detergent, 10
Synthetic granules process, 163

T

Tablets, 131
Tallow fatty alcohol, 11
Tetra-acetyl ethylene diamine (TAED), 7

W

Washing temperatures, 2, 6

Z

Zeolite- built formulations, 4

FSC
www.fsc.org
MIX
Papier | Fördert
gute Waldnutzung
FSC® C083411

Kolibri 360 GmbH
Ferdinand-Jühlke-Straße 7
99095 Erfurt, Deutschland
produktsicherheit@kolibri360.de